Life Between the Tides

Marine Plants and Animals of the Northeast

Les Watling, Jill Fegley, and John Moring

Illustrated by Andrea Sulzer

Edited by Susan K. White, Maine Sea Grant Program

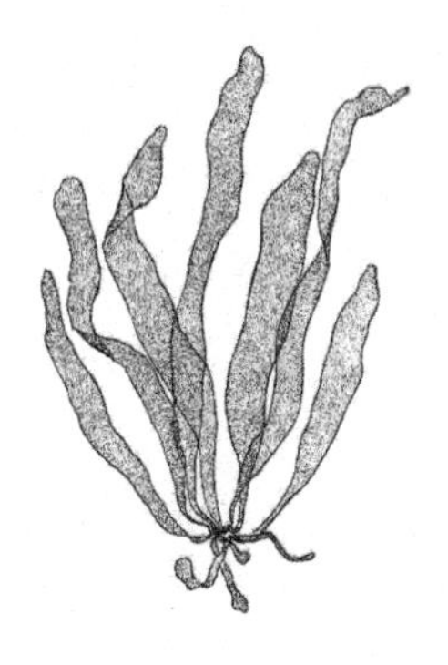

Tilbury House Publishers
ThomastON, Maine

Tilbury House Publishers
12 Starr St., Thomaston ME 04861
www.tilburyhouse.com
800-582-1899

First Edition May 2003

10 9 8 7 6 5 4 3

Library of Congress Cataloging-in-Publication Data
Life between the tides : marine plants and animals of the Northeast / edited by Susan K. White ; illustrated by Andrea Sulzer.— 1st ed.
p. cm.
ISBN: 978-0-88448-253-6
1. Intertidal ecology—Northeastern States. I. White, Susan K.

QH104.5.N58 L54 2003
577.69'9'0974—dc21

2002154600

Cover photograph by Jonathan Bird/www.oceanicresearch.org

Project Coordinators
Esperanza Stancioff, University of Maine Cooperative Extension/Sea Grant
Susan White, Maine Sea Grant

Writers
Les Watling, University of Maine
Jill Fegley, Maine Maritime Academy
John Moring, University of Maine

Reviewers
Invertebrates
Jim Carlton, Williams College-Mystic Seaport
Fishes
Grace Klein-MacPhee, University of Rhode Island
Plants and Seaweeds
Peg Van Patten, Connecticut Sea Grant

Dedication

Life Between the Tides is dedicated in memory of
John Moring,
professor of zoology at the University of Maine.
John was one of the few biologists who understood
the importance of the intertidal zone as a habitat for small fish.
John's love of wildlife inspired his teaching and enriched our lives.

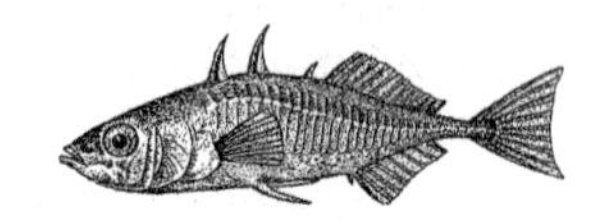

Table of Contents

ACKNOWLEDGMENTS

This book was inspired by the many coastal stewards involved in activities along the coast that help protect marine habitats and the organisms living there. We are especially indebted to the volunteer environmental monitoring community for giving us the idea for this guide in the first place.

Life Between the Tides is an outgrowth of *A Guide to Common Marine Organisms Along the Coast of Maine* published by Maine Sea Grant in 1998. Many individuals contributed to this first guide, including Laurie Bean, Susan Brawley, Ian Davison, MaJo Keleshian, Doug McNaught, Tim Miller, Wendy Norden, Robert Russell, Bob Steneck, Kate Sullivan, John Vavrinec, Lynn Wardwell, and Melissa Waterman. Their invaluable contributions helped make the first book a success.

For this book, we would like to thank the University of Maine's Darling Marine Center for providing holding tanks for the marine organisms used by the illustrator and the students who collected them.

Introduction

Most people think of Maine, northern New England, and eastern Canada as the region with the rocky coast. However, for those who know the shore well, it is clear that this region has not only its famous rocky shore, but also has an abundance of mudflats and salt marshes and, in the southern part of the region, many sandy beaches.

With such a diversity of habitats, it is not surprising that the northeastern U.S. and Canada is also home to a very large number of marine and estuarine species. Some, such as lobsters, crabs, and sea urchins, are familiar to everyone. Others, however, have not been seen by most people, and therefore are true novelties when observed for the first time. This guide was developed to help people find out what are the most common of these organisms and to learn something about how these species live their lives.

The book is divided into four sections. Chapter One gives a brief overview of the major habitat types to be found along the coast. Chapter Two introduces the diverse array of invertebrate animals. Invertebrates are all those animals that do not have a backbone, so this category excludes fish, reptiles, birds, and mammals. Chapter Three covers the small fishes likely to be encountered in tide pools and salt marshes; some are residents of the pools all year long, whereas others are merely the early juvenile stages of fish that spend most of their lives offshore. The larger plants of the marine fringe, including seaweeds, are covered in Chapter Four.

The first edition of this book, *A Guide to Common Marine Organisms Along the Coast of Maine*, was developed to help volunteer citizen monitors identify organisms encountered during water quality, phytoplankton, and habitat monitoring activities. However, it also proved to be popular with students and their teachers, as well as with local residents and other visitors to the shore. In this edition we have expanded the invertebrate chapter, rewritten the chapter on marine macroalgae (seaweeds) and included vascular marine plants, and added chapters on habitat characteristics and tide pool and salt marsh fishes. We hope that it will continue to be useful to all who appreciate coastal environments and want to learn more about this fascinating world.

Life Between the Tides

Marine Plants and Animals of the Northeast

Chapter One

Coastal Habitats

Les Watling, University of Maine

The beauty of the northern New England coast comes from its variety of habitats, often jumbled together but all within easy view. The habitats here differ from those found elsewhere along the coast of the U.S. For example, instead of the vast areas of salt marsh that exist in the mid-Atlantic region, Maine, for the most part, has small fringing marshes bordered by mudflats and rock outcrops. And, instead of the large shallow bays seen from New Jersey to North Carolina, northern New England has a diversity of bay sizes and shapes. This habitat diversity, however, is not only the key to the beauty of the region, but it is also essential to the high level of marine species diversity found in these waters.

Along the northeastern coast, the predominant habitat types to be found are: salt marsh, mudflat, sand flat, sandy beach, cobble and boulder beach, sheltered rocky shore, and exposed rocky shore. In some areas, several of these habitat types can be found within a few yards of each other but, especially in the southern part of the region, marshes may cover hundreds of acres and downeast of Mt. Desert Island, tidal flats may do the same.

Salt marshes are usually the last habitat to be covered by seawater at high tide. They are dominated by salt-tolerant grasses living in highly organic mud. These grasses bloom and produce seeds in mid- to late summer, just as do land-dwelling grasses. The highest tides often carry with them floating seaweed, wood, and human-made debris, which is deposited at the landward edge of the marsh in what is called the "wrack" line. Decomposition of the wrack material is aided by several species of crustaceans and insects that may also inhabit the adjacent marsh.

Mudflats extend from the salt marsh to the low tide line and are often separated from the marsh by a drop of 2 feet (0.5 meter) or so. While the upper sections of mudflats are almost always very muddy, and sometimes soupy, the lower reaches may have some sand mixed into the mud. The sediment (soil) is very rich in organic matter and watery, and it can get very hot on a warm summer afternoon. Decomposer bacteria love these conditions and their activities often result in mudflats being very "stinky." As a result, there is no oxygen in the sediment deeper than 1/32 in (1mm). Since all of the animals living in the mudflats need oxygen, they must build and maintain burrows or tubes into which they can pump oxygenated water when the mudflats are covered at high tide.

Sand flats develop where the supply of sand is high and the supply of mud is low. Usually these are areas where the flat is exposed to ocean waves during high tide. Because sand flats exist in "high energy" areas, muddy sediment and organic particles are not deposited in high amounts. Therefore, the sediment is coarse and does not hold water as well as muddy sediment does. As a result, sand flats do not have the high numbers of decomposer bacteria; they are "cleaner" and not very smelly. On the other hand, life for sand-flat-dwelling animals is a bit more difficult because food is harder to find within the sediment. Many sand-flat dwellers build tubes and most obtain their food from the overlying water at high tide.

Sand beaches, as well as cobble and boulder shores, exist in areas where the wave energy is even higher than it is on sand flats. Fine sedi-

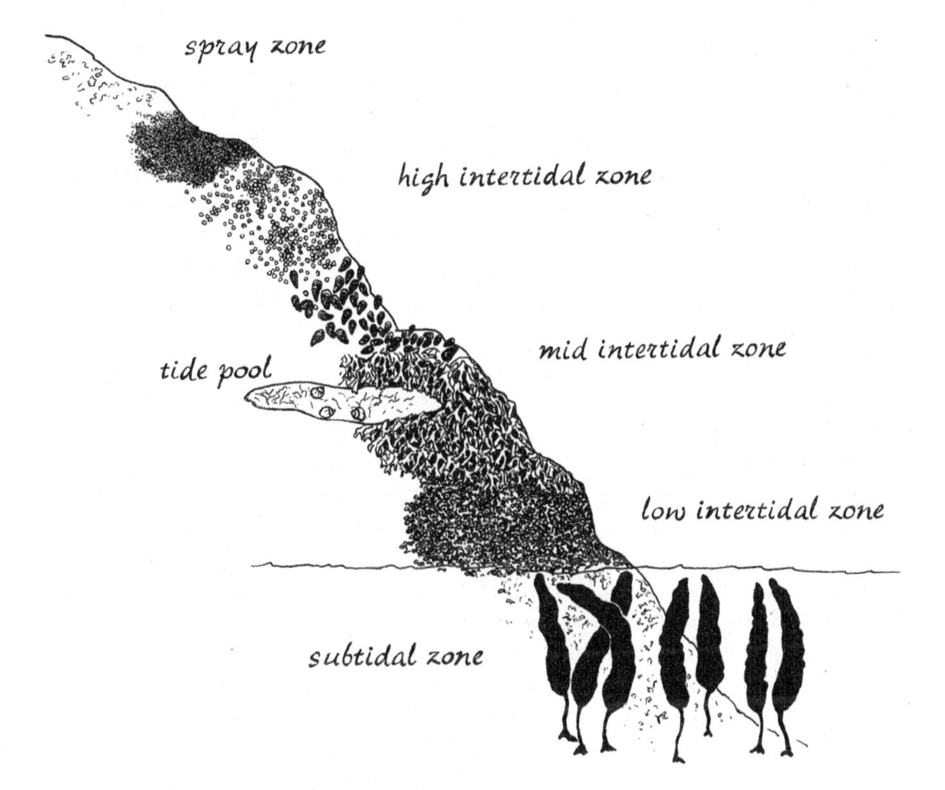

ment particles and organic matter do not stay long in these areas. Because of the increasingly larger sizes of the sand and pebbles making up these habitats, water drains through them readily at low tide and flushes the spaces between the sediment grains at high tide. Most of the animals in these habitats are out of sight, living in the micro-environments provided by the spaces between the sediment particles. Larger animals also spend most of their time buried in the sediment and, on sandy beaches, may migrate up and down the beach with the tide. In cobble areas, some animals more characteristic of rocky shores may be found attached to the largest boulders, but most of them can be found if you lift boulders and dig deep into the gravel.

An enduring feature of the northern New England coast is its rocky shoreline. Rock outcrops can be found almost everywhere, from the high wave energy, exposed outer coast to more sheltered locations far up in the bays and estuaries. The diversity of species living on rocky ledges varies with the wave energy. On the outermost rocks exposed to the full force of incoming storm waves, there may be only a few tenacious individuals. With just a small amount of shelter, however, diversity increases dramatically. Offshore rocks, kelp beds, changes in the slope of the rock face, or many other features may provide this shelter. Within the beds of red algae that dominate the lower reaches of the rocky shore, wave energy is also decreased, and very high numbers of typically sediment-inhabiting species may be found in the shell and sand particles trapped there. As a result, outer coast rocky habitats are areas prized by school groups and university biology classes for observing the range of animal and plant forms the ocean environment can support.

Wave energy decreases landward of the headlands where rocky ledges are also found. These habitats are dominated by brown algae, which drape the rocks and retain moisture, allowing other species to survive during low tide periods. Because the energy in these environments is lower than on the outer shore, the species composition is different and the overall levels of diversity are also lower.

All the habitat types described above have their characteristic species, many of which are restricted to only one habitat. Collecting animals or plants from all of these habitats should be done in moderation, with careful consideration given to habitat disruption. Walking on mudflats leaves large depressions (footprints), which may last for months—to say nothing of body imprints left when novices fall over, leaving their boots behind! Repeated tramping on the same path over a marsh will result in the grass dying along the path with subsequent colonization of the bare sediment by blue-green algae. Clambering over rocks in the mid-intertidal zone late in the summer or fall will almost always result in the dislodgment of barnacles, many of which have grown so vigorously throughout the summer that their attachments to the rock are, at best, tenuous. When rocks or cobbles are turned over to view the inhabitants underneath, they must be returned to their normal upright position, or many unseen species will die from exposure and desiccation.

Chapter Two

INVERTEBRATES OF THE SHORE

Les Watling, University of Maine

Marine invertebrates are very diverse, both in terms of numbers of species and ways in which their bodies are designed. To make identification easier, the common invertebrate species in this book are grouped by general body form. This arrangement does not imply that the invertebrates are related in an evolutionary sense but, instead, shows some common adaptations of invertebrates for living in particular habitats.

These are animals that are quite simple in their body design. With the exception of the swimming jellyfish, all lead a sedentary life and depend on water movement to bring them food and to disperse their young.

BLOBS, BAG-BODIES, AND COLONIES

Among the simplest of all animals are the **sponges**. While the cells of sponges work together, there is very little organization. Sponges use specialized cells to make a primitive form of inner "skin" that pumps water, from which they take very small food particles, and to make the spicules (little spikes containing

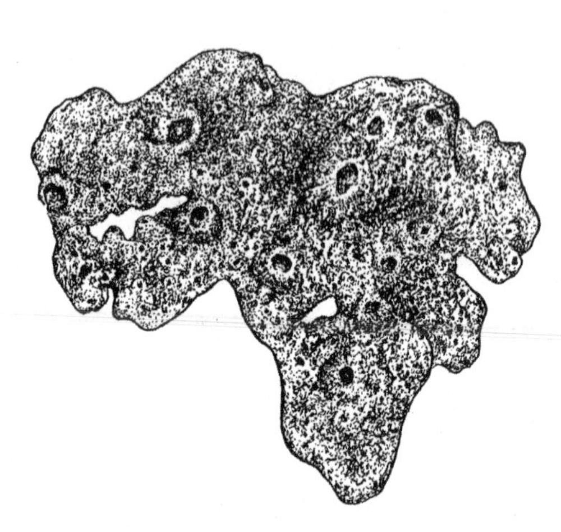

Bread crumb sponge *(Halichondria panicea)*—up to 2 ft (60 cm); in a crust, 0.5 in (1 cm) or so thick

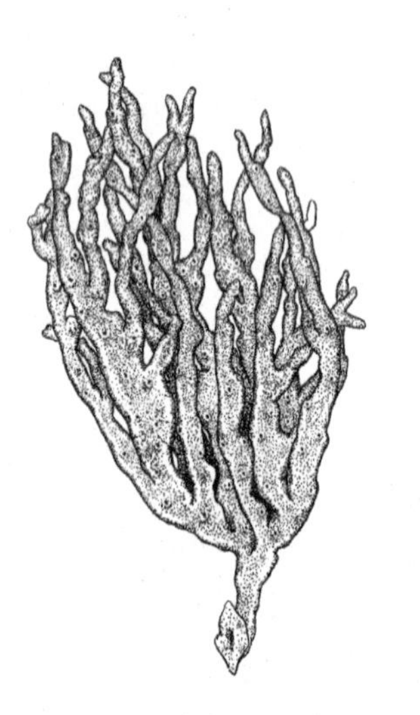

Eyed finger sponge *(Haliclona oculata)*—up to 10 in. (24 cm) high

silica or calcium carbonate) of the skeleton. Sponges live in or on all kinds of substrates, including mud, rock, and shells. Although sponges were once thought to be not very diverse, about fifty species now have been found in the Damariscotta River estuary (in midcoast Maine) alone.

Along the outer coast, the most commonly encountered sponge is *Halichondria panicea.* This species can be almost any color, although a dull greenish gray is most common. The green color comes from a microscopic alga that lives with the sponge colony and provides it with some food, presumably in return for the shelter the sponge provides. This sponge, often called the **bread crumb sponge** because of the way it breaks when handled, usually grows as an encrusting sheet on rocks in crevices or deep under overhangs. It also produces several sulfur compounds that give it a rather obnoxious smell.

A deeper-dwelling sponge, but the one most frequently found washed up on sandy beaches, is *Haliclona oculata*, also known as the **eyed finger sponge.** The colony consists of long, finger-like extensions with regularly spaced holes on the surface. These holes, called oscula, are where the water exits the sponge after it has been pumped through the outer cell layer and microscopic food particles removed. *H. oculata* lives subtidally, but it is often uprooted during storms. When alive, it is a soft orange-red color, but turns white soon after death.

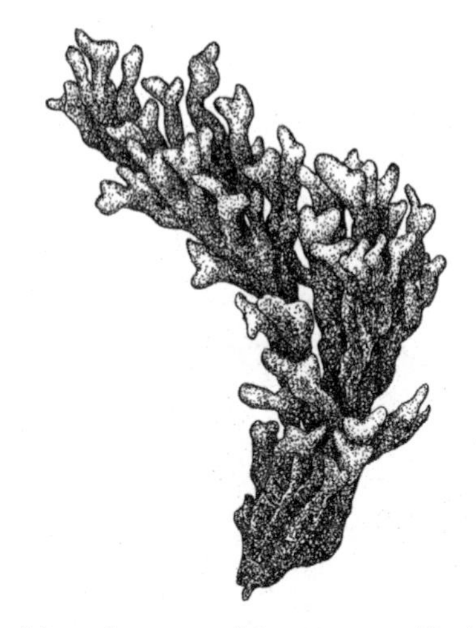

Red beard sponge *(Microciona prolifera)*—up to 10 in. (24 cm) in diameter

In the upper, warmer reaches of Maine's estuaries, the **red beard sponge**, *Microciona prolifera,* can be found. This sponge gets its name from its bright red color and scraggy form. It usually grows attached to rocks but, as the colony enlarges, it tends to trap mud particles; when quite large, it can be home to many smaller species of invertebrates. This species is more common south of Maine and is thought to be a relic of times past when Maine's coastal waters were much warmer than they are today.

Sulfur or boring sponge *(Cliona* spp.*)*—makes holes 0.1 in (1-2 mm) in diameter in shells; if massive, can cover entire shell

If a shell seems to have a large number of very small holes peppering its surface, chances are it is, or was, home to the **sulfur or boring sponge** of the genus *Cliona.* These sponges chemically burrow into oyster, clam, and mussel shells. When the sponge is young, the shell appears to be covered with small yellow bumps but, when allowed to grow to maturity, *Cliona* species may completely overgrow the shell,

Frilled anemone *(Metridium senile)*—often small (4 in or 7 cm) in intertidal zone, but can reach 18 in (45 cm)

forming a large yellow mass 10 cm (4 in) or more in size.

Anemones and their relatives belong to the phylum Coelenterata (which means "bag body"), also called the Cnidaria, due to the peculiar small, specialized stinging cells (known as *cnidae*, but also called nematocysts) located in their tentacles. Cnidarians may exist as polyps attached to the sea bottom, jellyfish floating freely in the sea, or have life cycles that include both stages. Anemones are found only as polyps; large jellyfish may have small, inconspicuous polyps; and in hydroids, both stages are often important.

Metridium senile, called **the frilled anemone**, is the most common anemone to be found on Maine's shores. It lives from the lower intertidal to about 98 ft (30 m) deep and may be green, white, orange, or various shades of brown. Intertidally, *M. senile* can be found clinging to the undersides of rocks where the humidity remains high at low tide. When disturbed, the anemone pulls its numerous fine tentacles into its mouth, resembling a fleshy lump on the rock. Anemones feed on small organisms, which they catch using the stinging cells in their tentacles.

Hydroids are very abundant along the Maine coast, especially in areas where there are rocks or other hard substrates to which the animals can attach. Hydroids are almost always colonial animals; that is, several individuals live together, attached by a common root-like system that enables them to share digested food materials.

An abundant species attached to floating docks along the Maine coast is the **pink hydroid** *Ectopleura* sp. Each part of the colony consists of a long stalk, at the end of which is a "head" bearing the food-gathering tentacles, a mouth, and a collection of reproductive structures.

Snail shells inhabited by hermit crabs are often partly overgrown by a bright pink, furry carpet called **snail fur**. This is a hydroid of the genus *Hydractinia*. It was long thought there was only a single species in Maine waters, *Hydractinia echinata*, but recent research shows that each hermit crab species has its own *Hydractinia* species associated with it. *H. echinata* is not found in Maine; instead, *H. symbiolongicarpus* is found on snail shells inhabited by *Pagurus longicarpus*, and *H. polyclina* is associated with the crab *P. acadianus*.

Pink hydroid *(Ectopleura* sp.)—2 in (5 cm)

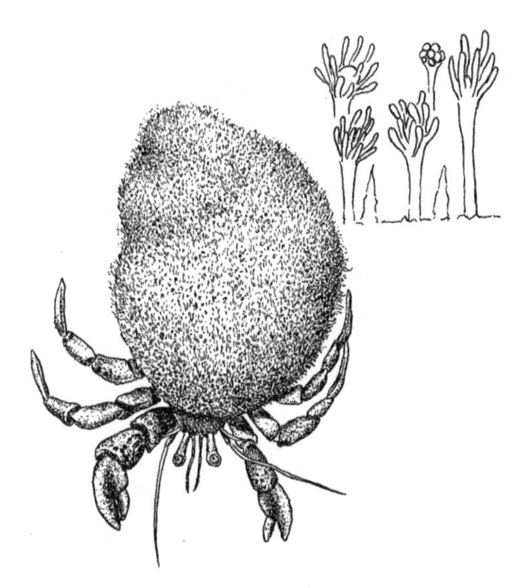

Snail fur *(Hydractinia* spp.)—0.1 in (1-2 mm) on shell inhabited by an Acadian hermit crab *(Pagarus acadianus)*; enlarged view above right.

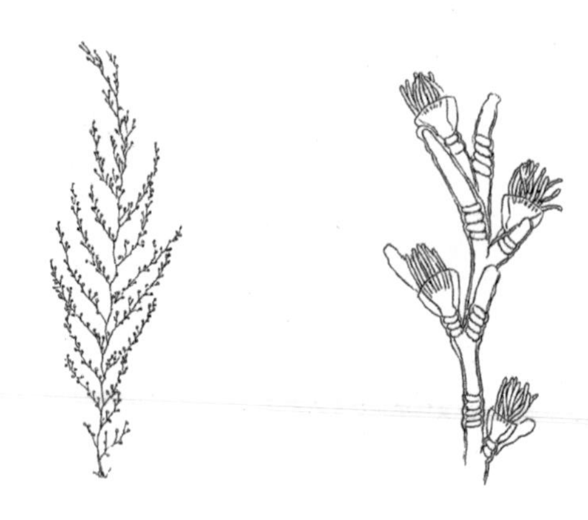

Campanularian or wine glass hydroids—generally less than 1.5 in (3 cm)

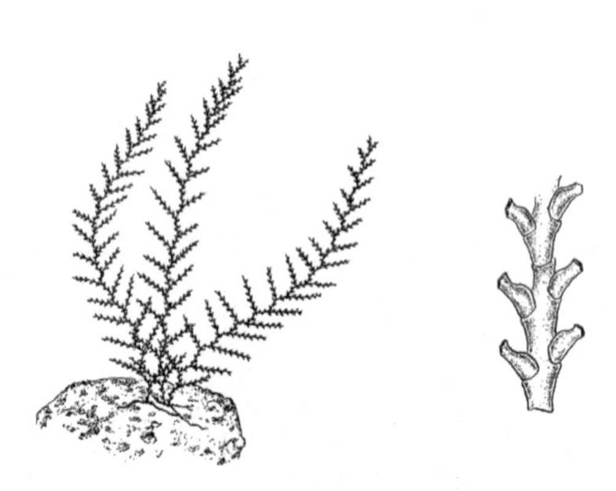

Sertularian or fern garland hydroids—often 2 in (5 cm) or large

Two other common groups of hydroid colonies are often seen in Maine. These are the campanularians (the **wine glass hydroids**) and the sertularians (the **fern garland hydroids**). Campanularians get their name from the small, wine glass-shaped cup in which the polyp sits. This type of hydroid is quite common on docks, rocks, shells, and plants, especially in shallow bays. Sertularians are so-called because the general form of the colony often looks like a fern frond. These hydroid colonies can be quite large and are often found growing on kelp and tunicate stalks, as well as rocks and other hard substrates. They are also more common along the outer coast, rather than in shallow bays and estuaries. Campanularians and sertularians differ in another important respect. In campanularians, the cup in which the polyp sits is open at the top but in sertularians, the cup has a lid, which can cover and therefore protect the retracted polyp. This difference most likely is the reason that sertularians can be found in abundance on the undersides of rocks, out of tide pools, along the outer coast.

Two species of true **jellyfish** commonly are found in Maine waters. The largest is the burgundy-colored **lion's mane jellyfish**, *Cyanea capil-*

lata, a species native to the colder waters of the Atlantic Ocean. Groups of robust tentacles are arranged along the margin of the thick, disc-like bell. These tentacles bear numerous stinging cells and are used to capture food. Prey items subdued by the tentacles are transferred to the mouth, located at the bottom of a long, ruffled tube hanging down from the center of the bell. Some people have a strong reaction to the toxin in the stinging cells of this jellyfish, so it should be handled cautiously.

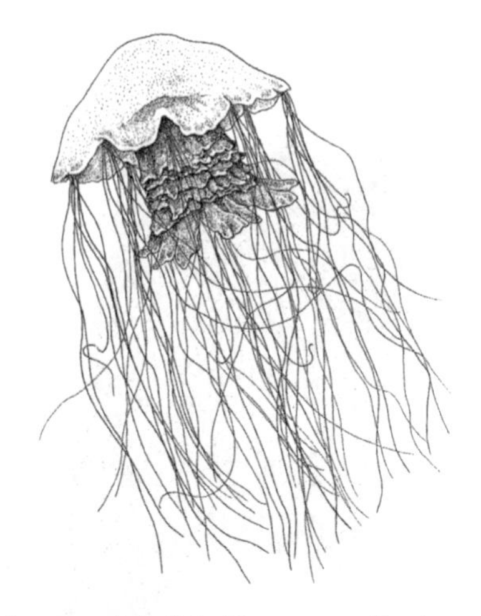

Lion's mane jellyfish *(Cyanea capillata)*—usually 12 in (30 cm) but may be larger

The other local jellyfish is the **moon jellyfish**, *Aurelia aurita*. It is nearly all white, or slightly transparent, except for the four horseshoe-shaped, light pinkish gonads located in the bell. This species has numerous short tentacles arranged along the edge of the bell. The stinging cells of the moon jellyfish are not toxic to humans, so this species can be handled easily.

A relative of the cnidarians is the **sea grape or gooseberry**, *Pleurobrachia pileus*, a member of a group called **ctenophores**. By holding a ctenophore up to the light, the ctenes, or comb rows (which gives the group its name) can be seen by their iridescence. Ctenophores are a major consumer of smaller zooplankton,

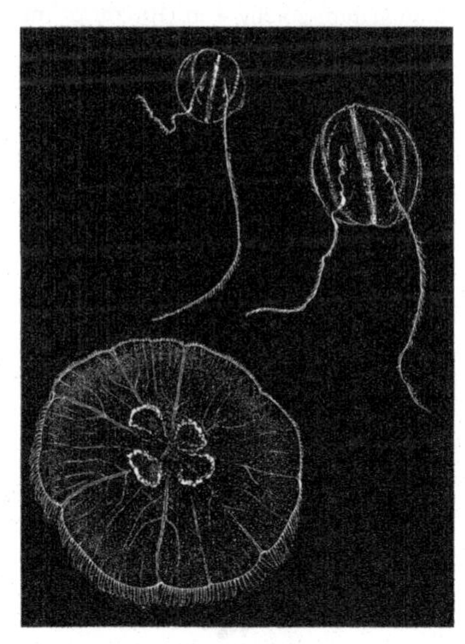

At top: **Sea grape or gooseberry** *(Pleurobrachia pileus)*—1.2 in (3.5 cm); and below, the **moon jelly** *(Aurelia aurita)*—up to 16 in (40 cm)

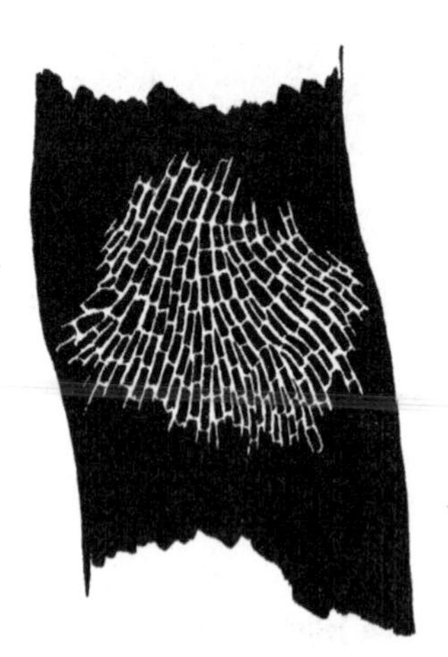

Coffin box bryozoan *(Membranipora membranacea),* enlarged view—individual zooids less than 1 mm; colony 4 in (10 cm) or more across. Below, colony on kelp blade.

especially copepods, fish eggs, and larvae. Prey are captured by adhesive, rather than stinging, cells located on the two long tentacles which can be retracted into pockets in the body.

Bryozoans are another group of colonial animals. Each individual makes its own box-like shelter, which is joined together with others in a complex pattern. Each individual is also the descendant of an individual next to it, and all individuals are connected by a small strand of tissue extending through the walls of the box. Bryozoans feed by using cilia on their tentacles to generate water currents, which carry very small particles into the vicinity of the tentacles where they are trapped by sticky substances. Because it is not efficient for the whole colony to feed at one time, cooperating individuals set up feeding zones.

There are many bryozoans in the Gulf of Maine region, most of which can be identified only with the aid of a microscope. A typical species is in the genus *Electra* (not shown), which forms small, lacy colonies on rocks and seaweeds. The introduced species, **coffin box bryozoan**, *Membranipora membranacea,* forms large, flat colonies on kelp blades. These colonies give the kelp blade a lacy-looking covering but, when the blade is touched, it is found to be rather hard. The hardness comes from the calcium carbonate that is secreted into the walls of the individual boxes.

Some bryozoans have soft box walls and are firm, rather than hard, to the touch. A common species of this type is the **bristly bryozoan**, *Flustrellidra hispida.* Along the edge of the box are many long spines which can be seen easily with the unaided eye, especially when held up to bright sunlight. *F. hispida* is found most frequently to the east of Penobscot Bay, often growing on kelp stalks, rocks, or pebbles. It usually has a brownish appearance, and may grow in flat encrustations or produce large, erect branches.

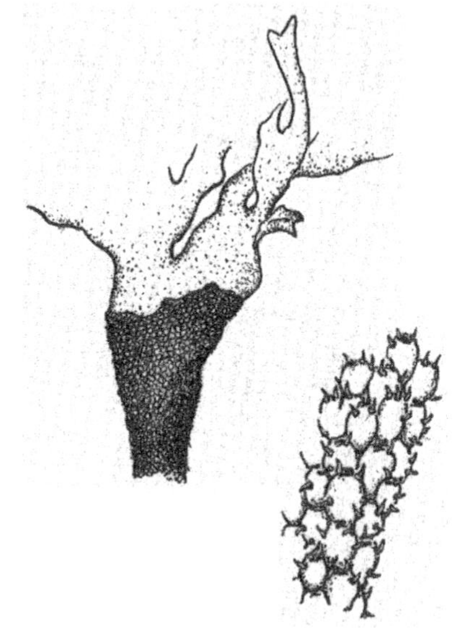

Bristly bryozoan *(Flustrellidra hispida)*—colony often 4 in (10 cm) high

Tunicates are sessile organisms that live either solitarily or in colonies. They are related to fish and other vertebrates, due to their tadpole-like larvae. Tunicates get their name from the tough, often leathery, outer tunics in which they live. Whether solitary or colonial, all tunicates have elaborate filter baskets that are used to extract very small food particles from the water. There are several tunicate species in Maine, including the three following species, which might be commonly encountered:

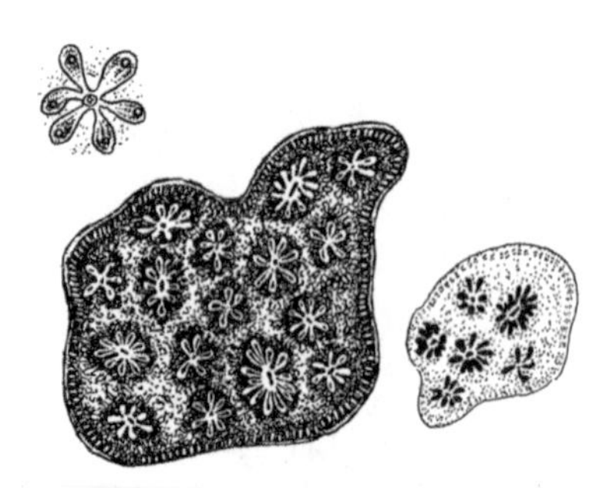

Golden star tunicate *(Botryllus schlosseri)*—colony usually less than 4 in (10 cm) across

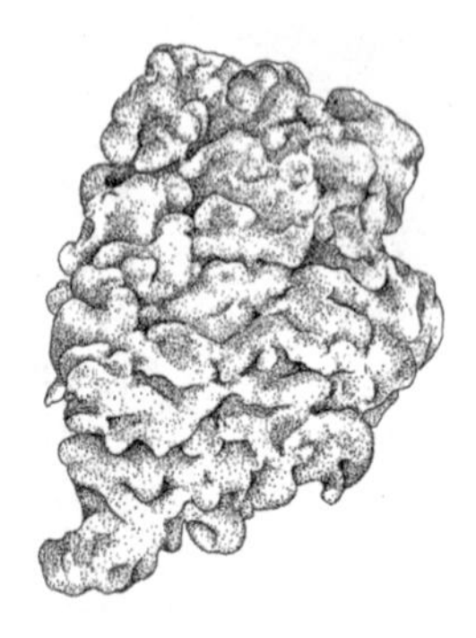

Orange tunicate *(Lissoclinum aureum)*—colonies very large, often exceeding 8 in (20 cm)

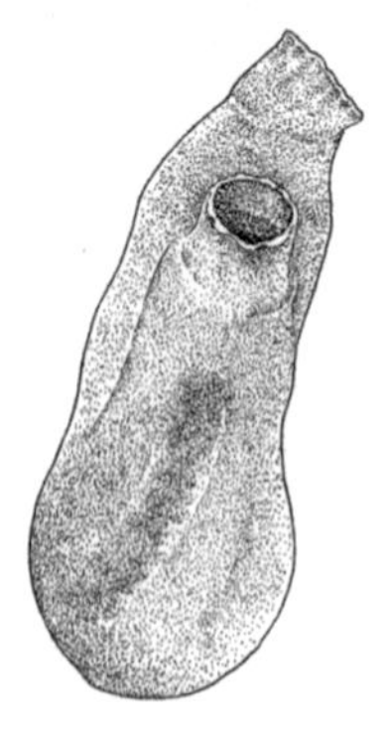

Sea vase *(Ciona intestinalis)*—up to 6 in (15 cm)

Golden star tunicate, *Botryllus schlosseri,* is a small, dark-colored colonial tunicate whose individuals have distinct white regions and are arranged in a multi-pointed star shape. There is a related species, *Botrylloides violaceus,* which is very similar in appearance except the individuals are not in a star-like arrangement.

Another common colonial tunicate is the **orange tunicate**, *Lissoclinum aureum.* It forms large, sponge-like colonies that often are found on pilings, eelgrass, or any hard substratum to which they can attach. The colonies may be white, purple, or orange.

Among the solitary tunicates, the vase-like, nearly transparent *Ciona intestinalis* is especially striking. As with all tunicates, each individual **sea vase** has two large openings—one where water is pumped into the filter basket area and the other where the water exits. This species can be found most easily by looking on pilings or in floating lobster pens.

There are also several other solitary tunicates that are usually small, brown, and nearly globular. These are members of the genus *Molgula*, commonly called sea grapes. It is difficult to identify the species in this genus because they differ from one another only in small details, such as the pattern on the outside of the filter basket, and the length and direction of the loop in the intestine. **Sea grapes** are quite common in the brackish parts of New England bays and can be found growing on pilings, sea grasses, rocks, or any other moderately solid surface.

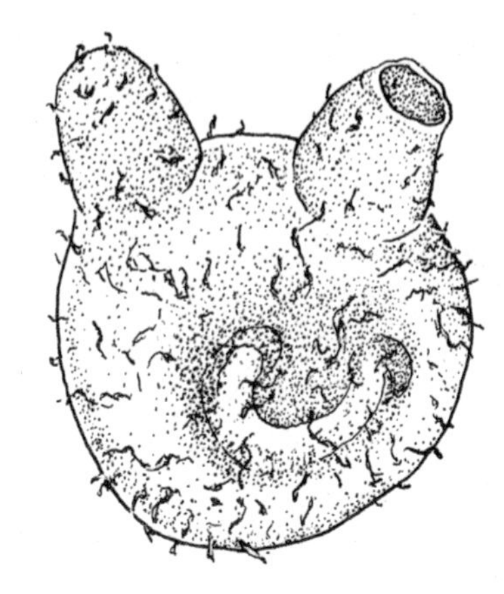

Sea grapes *(Molgula* spp.*)*—usually about 0.8 in (2 cm)

The largest of all the tunicates likely to be found near shore, or washed up on the shore after a storm, is the **sea potato,** *Boltenia ovifera.* The tunic of this tunicate is extended into a long stalk that is attached to a rock or pebble. A large number of hydroids and bryozoans are usually attached to the stalk. This tunicate is a very large species; the stalk is often 4 to 8 in (10 to 20 cm) long and the body may be 1.6 to 3 in (4 to 8 cm) long.

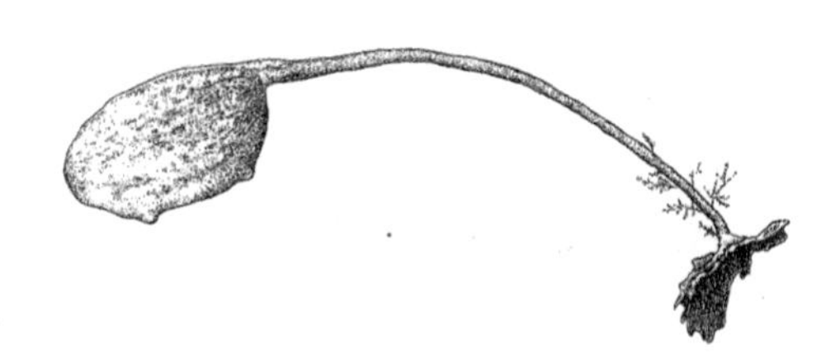

Sea potato *(Boltenia ovifera)*—body about 3 in (8 cm); stalk as high as 12 in (30 cm)

Worms and Worm-Like Creatures

Worms may be a variety of shapes and sizes. In the sea, the true worms are called polychaetes, named after the numerous chaetae, or bristles, seen along the sides of their bodies. Other worm groups include the flatworms and ribbon worms. Flatworms are unsegmented, very flat, and glide over the substrate on a layer of secreted slime, using cilia for propulsion. Although ribbon worms are also usually flat, they are generally quite long relative to their width and may be colored dark green, brown, or bright pink. They use both muscular contraction and cilia, in a slime layer, for movement.

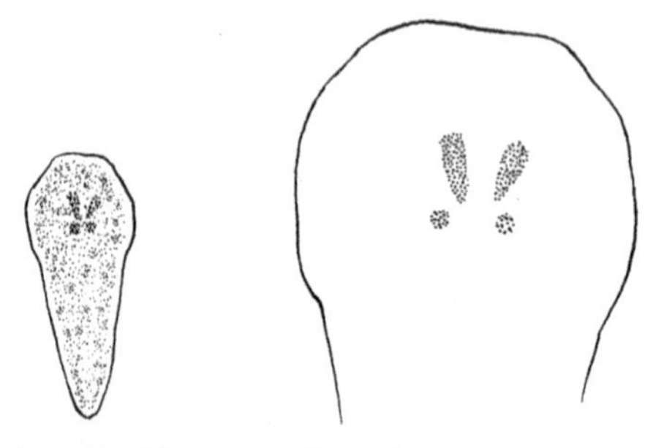

Speckled flatworm *(Notoplana atomata)*—0.8 in (2 cm)

Most of the **flatworms** found in Maine are small and inconspicuous. Two species, however, can be found readily in intertidal areas. *Notoplana atomata,* the **speckled flatworm,** lives on the undersides of rocks and is especially abundant in areas where large rocks lie on top of gravel or broken shells. It is a predator, trapping small animals in the sticky slime it uses for movement. The animal engulfs prey with its proboscis that it protrudes from its ventral side. *N. atomata* is a light gray to tan color, is about 1 in (3 cm) in length, and has a large number of very small, black eyespots on the upper surface and along the edge of the body.

The other common flatworm species is the **oyster flatworm**, *Stylochus ellipticus*. It is also a light gray to tan color, has eight to twelve eyespots on the upper surface and two small antennae which extend upward from the anterior part of the body. This species can be found in the same habitats as *N. atomata*, but also occurs in dead mussel shells and other muddier areas.

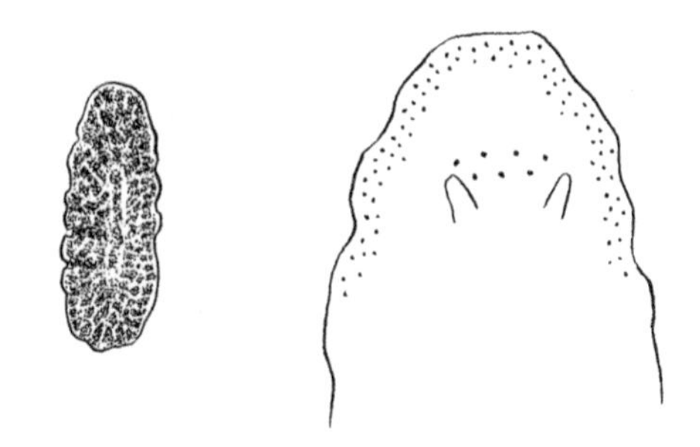

Oyster flatworm *(Stylochus ellipticus)*—1 in (2.5 cm)

Ribbon worms (in the phylum Nemertea) are the premier predators of muddy and sandy habitats in Maine. Most are dark colored but some, such as the **milky ribbon worm**, *Cerebratulus lacteus*, may be bright pink. Nemerteans use a long proboscis to catch their prey. This proboscis is stored in the upper part of the body and can be everted, using a mechanism much like extending the finger of a glove after it has been pushed inward from the end. The proboscis may be sticky or have a small spike at its end. Usually the proboscis is wrapped around the prey (which may be one of the segmented worms or a small crustacean) and some digestive enzymes are secreted onto the prey's body. The prey may then be pulled into the nemertean's mouth. *Cerebratulus lacteus* can grow to be almost 3 ft (1 meter) in length. It extends its proboscis into the open-

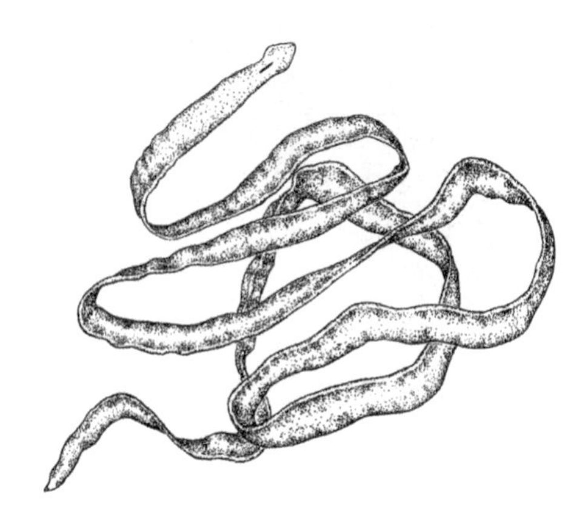

Milky ribbon worm *(Cerebratulus lacteus)*—up to 23 in (60 cm)

ing of a soft-shell clam and digests the clam from the inside.

The segmented worms, including the **polychaetes and oligochaetes**, dominate the muddy and sandy habitats of the Gulf of Maine. Oligochaetes (the group to which the familiar earthworms belong) are long, thin, thread-like, and often bright red in color. They are most often found living in large masses under stones along rocky shores. The taxonomy of marine oligochaetes from our waters is not well known, so not much more can be said about them. The polychaetes, on the other hand, are quite well known and are very diverse in our area.

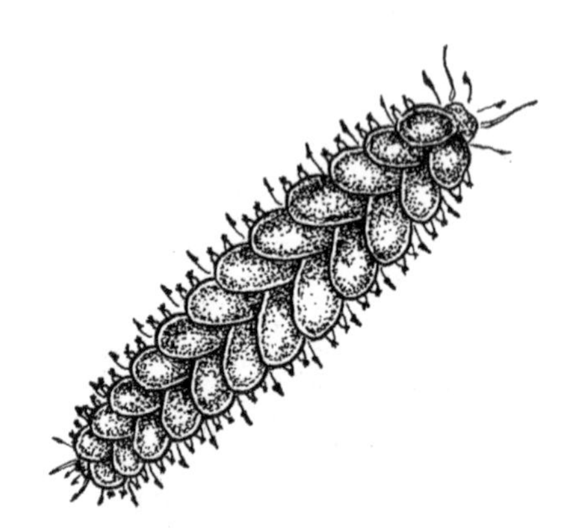

Twelve-scaled worm *(Lepidonotus squamatus)*—2 in (5 cm)

One group of polychaetes, vaguely reminiscent of large insect larvae, are the **scale worms**, so-called because their entire upper surfaces are covered with pairs of overlapping scales. *Lepidonotus squamatus* has twelve to thirteen pairs of scales, whereas several other Gulf of Maine species have fifteen pairs, at least as adults. *L. squamatus* commonly can be found in old mussel shells and under debris in the lower intertidal zone—from the rocky shore out to mussel reefs at the lower end of tidal flats.

The most commonly encountered polychaetes in New England waters, clam worms and bloodworms, form the basis of the baitworm industry. **Clam or rag worms** are members of the genus *Nereis*. Five species can be found, the most common being *N. virens* and *N. diversicolor*. All have complex heads bearing fleshy lobes (called palps) on either side of their mouths, four pairs of finger-like appendages (known as cirri), one pair of antennae at the front, and four eyes. Clam worms are predators, capturing their prey with a proboscis, turned inside out through the mouth and armed with strong jaws. They are also quite beautiful with iridescent green and purple bodies.

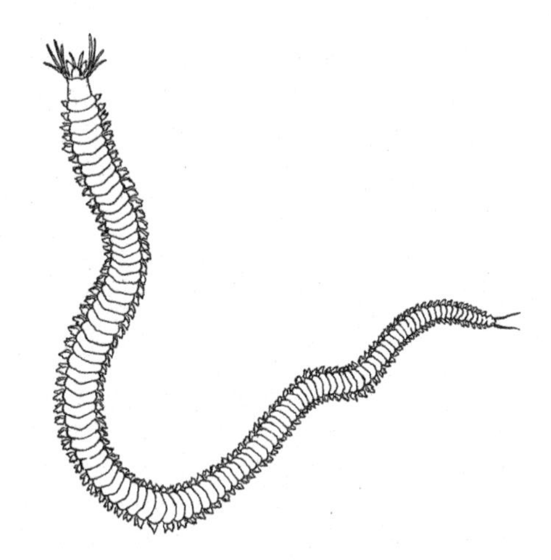

Clam or rag worms *(Nereis* spp.*)*—up to 8 in (20 cm)

The other major baitworm species is the **bloodworm**, *Glycera dibranchiata*. Bloodworms are easily distinguished from clam worms by their uniform red color and pointed heads. A bloodworm has no eyes or cirri on its head, which is often difficult to distinguish from its posterior end. However, if you handle a live bloodworm, you will soon know which end bears the head, as the worm will evert its long proboscis with little provocation. The proboscis bears a pair of sharp, black jaws that can pierce human skin. Bloodworms use a neurotoxin to

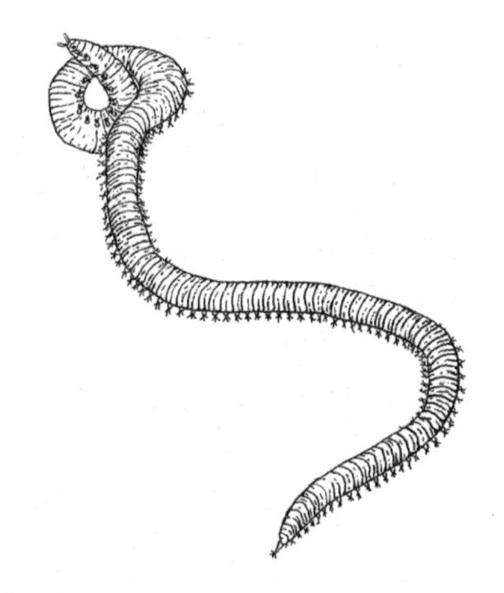

Bloodworm *(Glycera dibranchiata)*—up to 15 in (38 cm)

subdue their prey. It is injected into the prey through small pores in the bloodworm's jaws. This toxin sometimes causes mild allergic reactions in people, and the pinch of the jaws can be painful.

Mudflats are home to several species of worms, although finding them is not always easy. This is because most mudflat worms live in tubes or burrows and spend most of their lives within the mud. These species use the mud for food, as well as a place to live. In some cases, the presence of the worm can be detected by closely examining the mudflat surface for the fecal pellets that the worm leaves while feeding. Each worm species has its own fecal pellet design. Some are long and rod-shaped, others are the shape of footballs, and still others are coiled like ropes. Most are very small but can be seen when examining the mudflat surface from a distance of a foot or less.

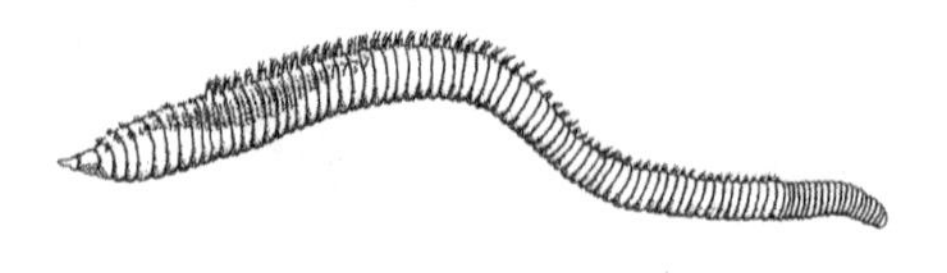

Leitoscoloplos spp.—usually less than 4 in (10 cm)

The most common mudflat polychaetes belong to the family Capitellidae. When digging in the mudflat with a shovel, one often sees a long, thin, very stretchable, wine-red colored worm. This is the **threadworm**, *Heteromastus filiformis* (not shown). It lives with its head deep in the mud (perhaps 10 cm or more), where it feeds, and it leaves a neat pile of very black football-shaped fecal pellets on the surface. Threadworms have no strongly developed appendages on the outsides of their bodies, which is a good indication that this type of worm spends all of its time burrowing into the mud in search of food. Several relatives of this species, especially *Mediomastus ambiseta* and *Capitella* sp., may also be found in Maine's mudflats, but they are difficult to distinguish without a microscope.

Other mudflat species that can be encountered quite commonly in Maine are in the genus ***Leitoscoloplos***. These animals are usually bright red in color and the posterior two-thirds of their bodies are covered with what appears to be a layer of straps extending over the back. These straps are the animal's gills, which it uses to obtain oxygen from the water when the tide is in. *Leitoscoloplos* species feed with their heads about 1.5 to 2 in (3 to 4 cm) deep in the sediment. This zone has no oxygen and usually has a buildup of poisonous hydrogen sulfide. By having the gills away from the head, the animal can find food in the anoxic zone, yet still get enough oxygen from the water coming over the mudflat.

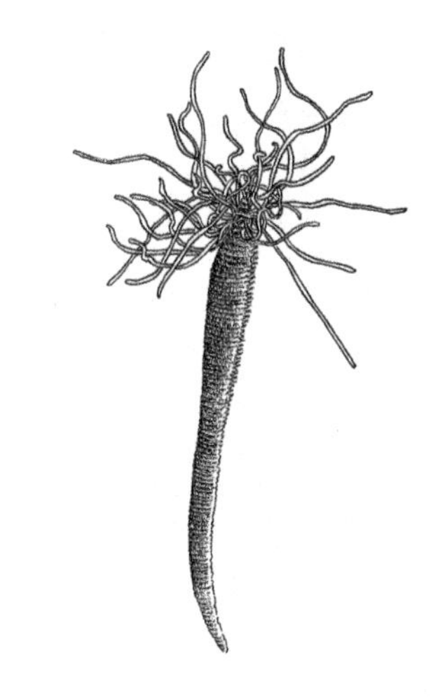

Fringed worm *(Tharyx sp.)*—0.6 in (1.5 cm)

Members of the family Cirratulidae use a similar feeding and respiring strategy. On Maine's mudflats, the most common cirratulid is a species of the genus *Tharyx*. Cirratulids are also called **fringed worms** because their bodies appear to have a fringe of fine threads along the length. This fringe, in fact, is the gills. Each thread is hollow and is flooded with hemoglobin-bearing blood, which gives the worm an overall bright red color. *Tharyx* spp. are encountered most often under the edges of stones or shells that lie on the mud surface. Usually there is a strong buildup of rotting organic matter in those places that the worm uses for food.

Clymenella torquata has one of the strangest lifestyles of all worms. It belongs to the family Maldanidae, also commonly referred to as

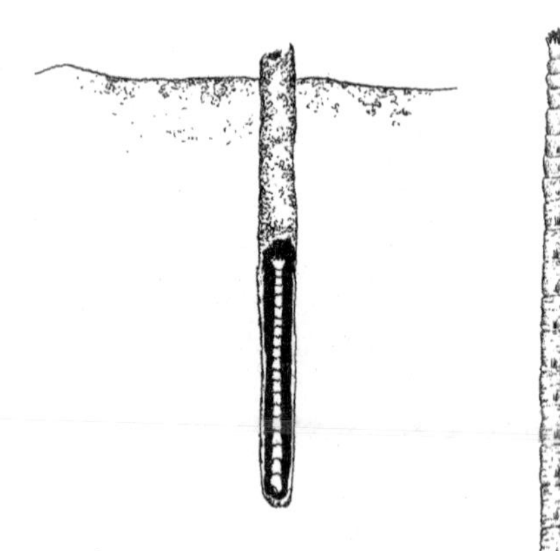

Bamboo worm *(Clymenella torquata) in its tube home (above), and at right*—4 in (10 cm) or less

bamboo worms because of their bright body colors and the odd fold in the skin near where their bodies are segmented. All bamboo worms have spade-shaped heads and crowns around their anuses. C. *torquata* builds a tube from sand grains and lives in the tube with its head at the bottom. It uses the spade-like top of its head to dig in the mud for food. Periodically, it backs up in its tube to defecate, which exposes its hind end to possible predation by fish. In any one mudflat, as many as 20 percent of the bamboo worms may be regenerating their rear ends.

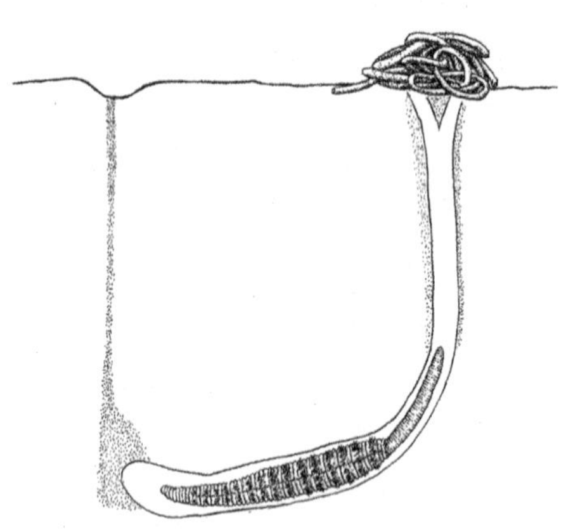

Northern lugworm *(Arenicola marina)*—up to 8 in (20 cm)

Feeding deep within the sediment is a strategy employed by many worms living on tidal flats, but that does not necessarily mean that the worm's food comes from those lower layers. The **lugworm**, *Arenicola marina*, has developed a special pumping strategy that allows it to feed on particles that were recently on the surface. Lugworms live on sandy tidal flats, residing in burrows that are 4 to 6 in (10 to 15 cm) deep. They produce long, coiled fecal pellets about the diameter of a pencil. The animal's feeding depression is about 8 in (20 cm) away from the pellet pile. *A. marina* pumps water through its burrow from the posterior end toward the head. The water pressure generated is used to stir the sand particles in front of its mouth, which are then ingested. As sand is moved through the gut, from the head to the fecal pellet mound, more sand and food particles fall into the depression and eventually make it to the feeding zone.

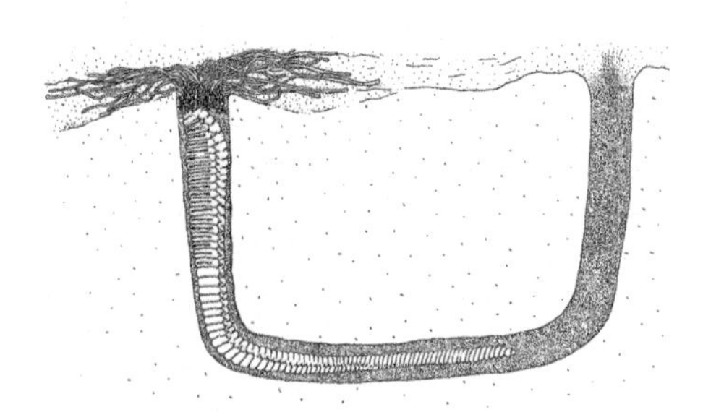

Ornate spaghetti worm *(Amphitrite ornata)*—up to 15 in (38 cm). Below, worm on mud surface, as seen from above.

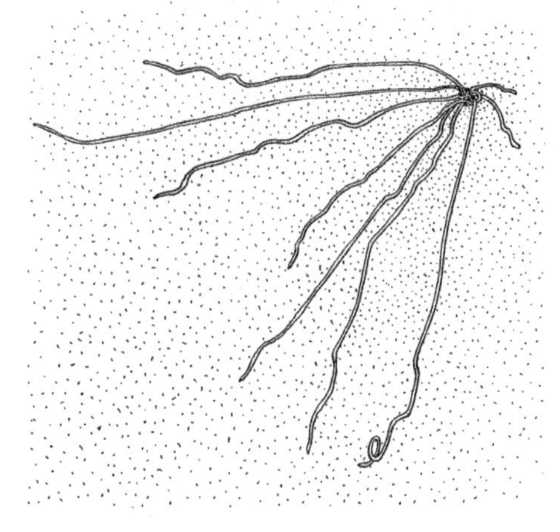

A mudflat species that also makes a thickly coiled fecal pellet and lives in a U-shaped tube is the Terebellidae species, *Amphitrite ornata*. However, rather than feeding deep in the mud, this animal uses very long, thin tentacles to gather its food from the surface of the mudflat. These tentacles, of which there may be as many as 100 or more, can stretch up to 20 in (0.5 meter) from the burrow, allowing larger worms to feed over very extensive areas. *A. ornata* is also known as the **ornate spaghetti worm** because it has a mass of frilly red gills and many long, white, somewhat flattened tentacles at its anterior end.

When looking closely at rockweeds or under rocks, one will often see small white spirals attached to the surfaces. These are the tubes of a small serpulid polychaete, the **spiral tubeworm,** usually one of the species of the genus *Spirorbis*. These worms have very small, sticky

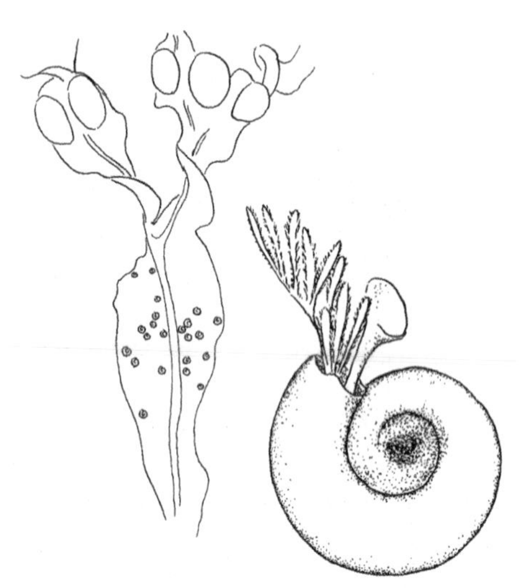

Sinistral spiral tubeworm *(Spirorbis borealis)*—0.1 in (3 mm). Left, tubeworm on rockweed.

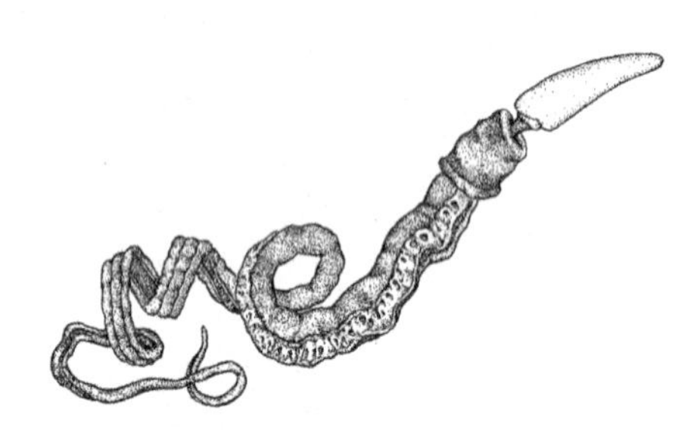

Acorn worm *(Saccoglossus bromophenolosus)*—less than 4 in (10 cm)

tentacles that they use to catch minute food particles suspended in the passing water. One of the tentacles is modified to form a plug, which can be used to close off the entrance to the tube after the **spiral tubeworm** has withdrawn into it. While it can be difficult to distinguish the various species, the most common species, *S. borealis*, coils to the left and has a smooth tube.

It may be surprising to learn that many of our relatives—that is, animals without backbones but with other features that put them early in our evolutionary line—are not designed like fish or lizards but, in fact, are sac-like (the tunicates) or worm-like. One of the worm-like animals is the **acorn worm**, which is very abundant on Maine's mudflats. An acorn worm has a milky white proboscis that is pushed into the mud to make the burrow and feeding gallery. Behind the proboscis is the collar, which contains the mouth. There are two species of acorn worms in Maine, *Saccoglossus kowalewskii* (not shown), ranging as far north as York, and *S. bromophenolosus* that can be found from York northward. These two species produce anti-bacterial chemicals in the lining of their burrows, and they were discovered to be different, based on the chemicals they produce.

Shell-Bearers and their Relatives

Mollusks are unsegmented animals with bodies generally covered by shells. Typical mollusks are clams, snails, tooth shells, and chitons. Some mollusks—such as squids, octopuses, and nudibranchs (which are snails)—have no shells.

Bivalves are mollusks with two shells, usually symmetrical left to right. Many bivalves live under the sediment and use special structures, called siphons, to pump water from above the sediment to their gills. Mussels and other bivalves that live on rocks or shells live up in the water, so their siphons are short or absent. Bivalves are unusual in another respect; their gills are used for respiration as well as for food collection. As the water passes over their gills, oxygen is passed to the blood, and very small food particles are trapped in a sticky mucus layer and transferred to their mouths.

Ribbed mussel *(Geukensia demissa)*—up to 4 in (10 cm)

Three species of mussels are commonly found in the Gulf of Maine—the ribbed mussel, *Geukensia demissa*; the blue mussel, *Mytilus edulis*; and the horse mussel, *Modiolus modiolus*. All occur in the intertidal zone and feed on tiny suspended particles.

The **ribbed mussel**, so-called because of the delicate lines running the length of the shell, is usually found living partially buried in the mud near salt-marsh grasses. It is a southern species, with Maine being near the northern edge of its range.

Blue mussel *(Mytilus edulis)*—up to 4 in (10 cm)

Blue mussels occur intertidally in bays and on the outer coast, but they are seen most commonly attached just below the water line of floats or other man-made structures in the water. On the outer coast, blue mussels are usually quite small and form a conspicuous band on the rocks between the barnacle and rockweed zones. In Maine, mussel "bars" are typically found near the low tide line of mudflats. They provide living space to a large number of other small invertebrates, thus enhancing the local biodiversity.

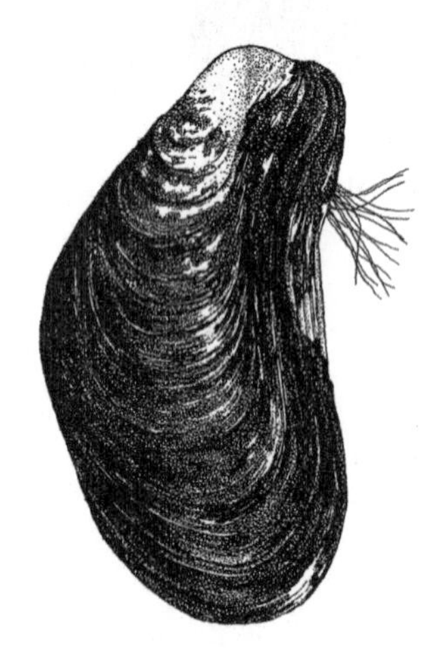

Northern horse mussel *(Modiolus modiolus)*—up to 6 in (16 cm)

The **horse mussel** is the largest of the mussels. It often has a shaggy appearance, due to the excessive amount of brown and papery outer shell layer, called the periostracum, which it produces. Horse mussels live in colder waters and are found from the low intertidal of the outer coast to about 98 ft (30 meters) deep. Near the entrance to the Bay of Fundy, there are large sub-tidal "reefs" made up entirely of horse mussels.

Several species of clams inhabit the Maine coast, but only a few are abundant in shallow, intertidal waters. Of course, the most notable clam species is the **soft-shell clam**, *Mya arenaria*, which is found on all mud flats from the Gulf of St. Lawrence to Delaware. Its ovoid shape, relatively soft shell, and united siphons covered with a brown "skin," distinguish it from other local species. While soft-shell clams live entirely under the surface of the mud, they use their siphons to obtain food and oxygen from the overlying waters when the tide is in.

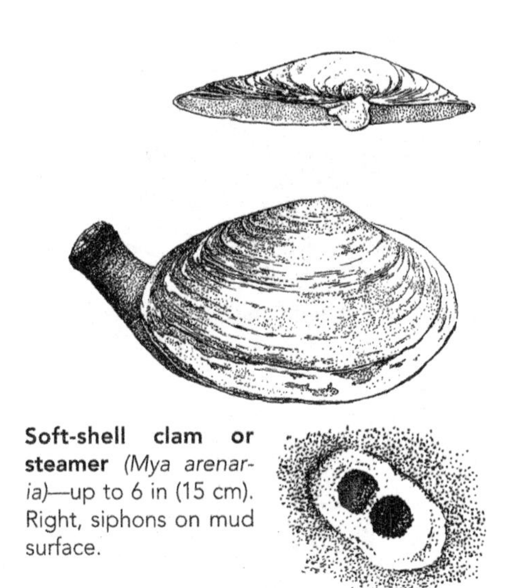

Soft-shell clam or steamer *(Mya arenaria)*—up to 6 in (15 cm). Right, siphons on mud surface.

The clam, *Macoma balthica*, is also found on tidal flats, usually living in exactly the same habitat as *Mya arenaria*. An easy way to identify the *Macoma* clam is by its siphons that are long and slender and not united. In fact, if you look closely at the surface of a mudflat, *Macoma* siphons can often be seen moving like little white worms in a circle, picking up bits of mud that are carried to gills for sorting and eating.

Macoma balthica—1.5 in (3.8 cm)

The **razor clam**, *Ensis directus*, is an inhabitant of sandy flats and lower edges of quiet sand beaches that is easily identified by its very elongate form and amber color. A razor clam can be difficult to catch because it maintains a burrow in the sand into

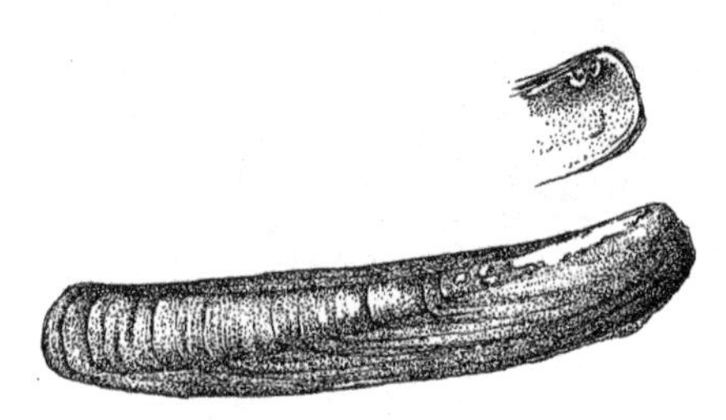

Razor or jackknife clam *(Ensis directus)*—up to 10 in (25 cm)

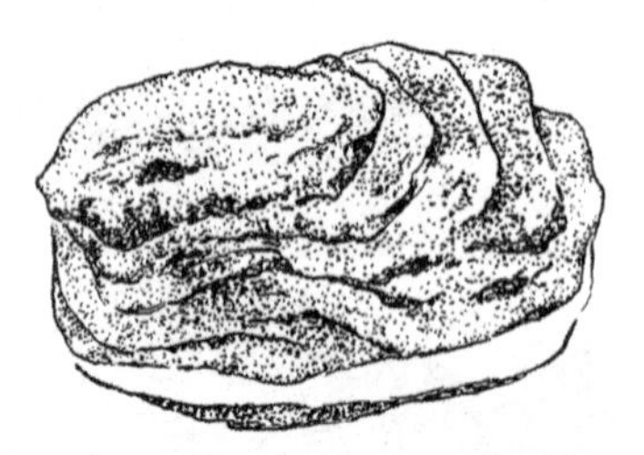

Arctic rock nestler *(Hiatella arctica)*—up to 1.5 in (3.8 cm

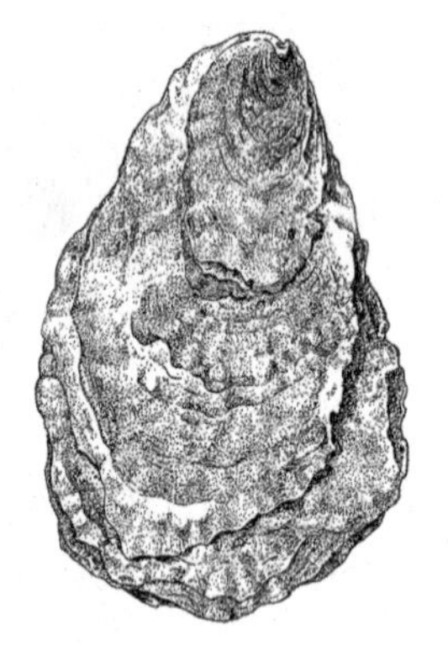

Eastern oyster *(Crassostrea virginica)*—up to 10 in (25 cm)

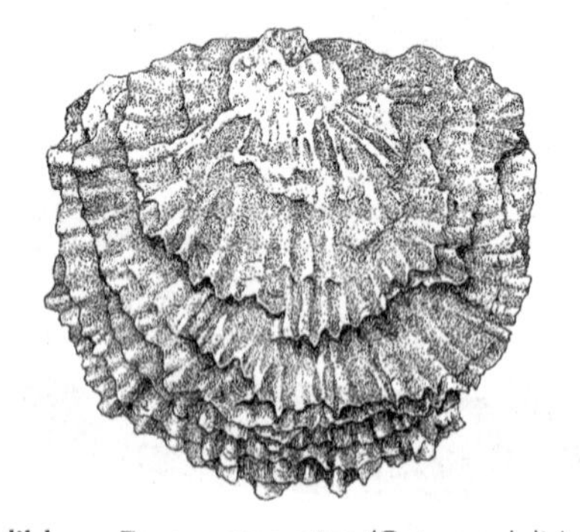

Edible or European oyster *(Ostrea edulis)*—up to 6 in (15 cm)

which it can retreat by rapidly contracting its foot. Contrary to popular belief, razor clams do not rapidly burrow into the sand when disturbed; they merely retreat quickly into a burrow that is already there.

Another small clam that is often overlooked in Maine is the **Arctic rock nestler**, *Hiatella arctica*. It is often irregular in shape due to its habit of living wedged under rocks, in crevices, under kelp holdfasts, and in other small spaces in the lower part of the rocky intertidal zone. This clam attaches itself by means of byssal threads and, if dislodged, cannot reattach itself since it does not readily make new byssal threads. Very young specimens are occasionally mistaken for young soft-shell clams.

Two species of oysters are likely to be found—the native **Eastern oyster**, *Crassostrea virginica*, and the **edible (or European) oyster**, *Ostrea edulis*, introduced from Europe. The Eastern oyster is more elongate in shape, whereas the edible oyster, especially when cultured, is more circular when viewed from above. Eastern oysters can be found in isolated pockets in the upper parts of estuaries in the southern part of Maine. Edible oysters, on the other hand, were brought to Maine to foster a fledgling aquaculture industry. Although scientists thought that Maine water temperatures were too cool in the summer for this species to reproduce, they have found occasional individuals in unusual places, suggesting that reproduction and survival may have occurred. There is evidence that *O. edulis*, has spread along the east coast of the U.S. as far south as Rhode Island.

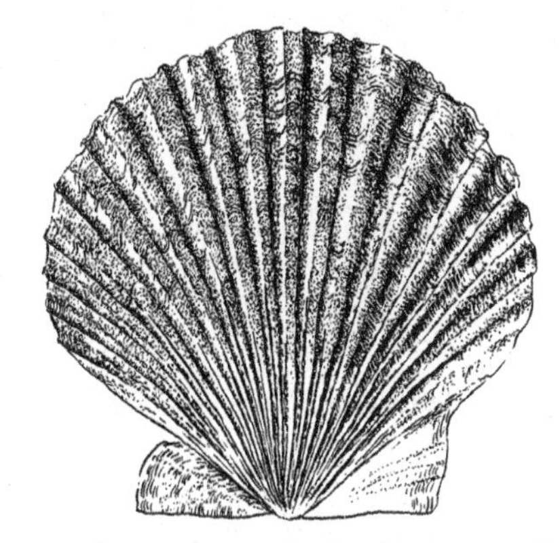

Bay scallop (*Argopecten irradians*)—3 in (7.5 cm)

The scallop shell is a familiar symbol of ocean life. In Maine, two scallop species are caught commercially. The smaller of the two is the **bay scallop**, *Argopecten irradians*, distinguished by its heavy ribs, which extend from the hinge side to the open edge. On the other hand, the **sea scallop**, *Placopecten magellanicus* is quite large and the shell is sculptured with numerous fine lines. Due to Maine's deep and cool bay waters, both species are likely to be found near shore.

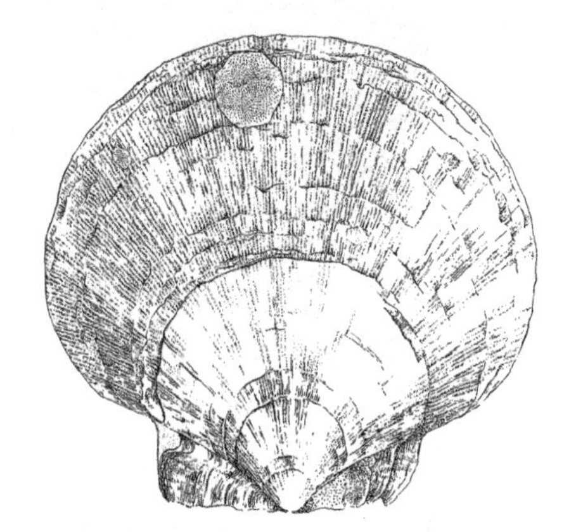

Sea scallop (*Placopecten magellanicus*)—up to 9 in (23 cm)

Another group of mollusks are the **chitons**. A chiton has a broad muscular foot, which it uses to hold its body tight against the surface of the rock. Eight calcareous plates, firmly embedded in a leathery mantle, protect its back. While a chiton may not move while being observed, it is quite mobile and glides slowly over

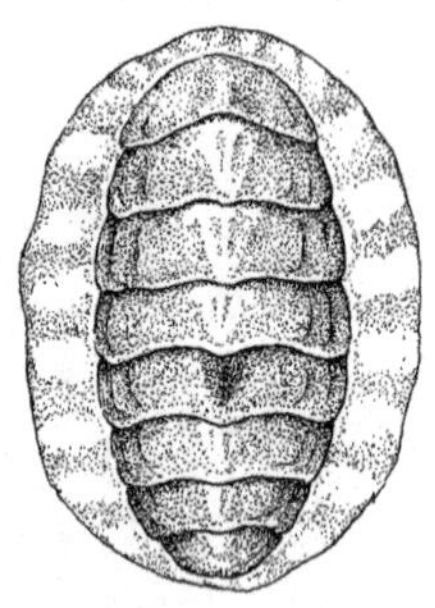

Northern red chiton *(Tonicella rubra)*—up to 1 in (2.5 cm)

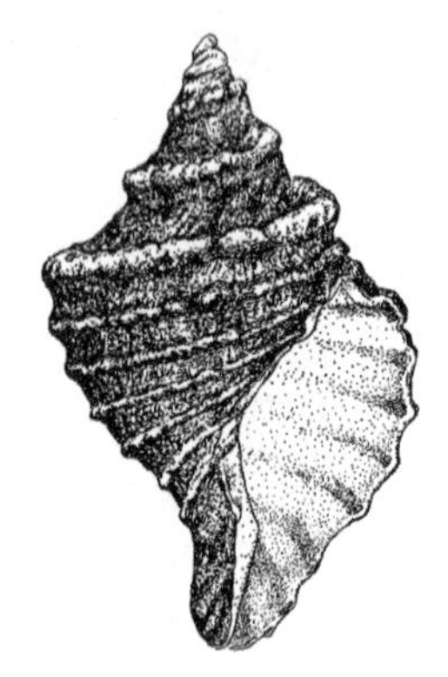

Ten-ridged whelk *(Neptunea lyrata decemcostata)*—up to 5 in (13 cm)

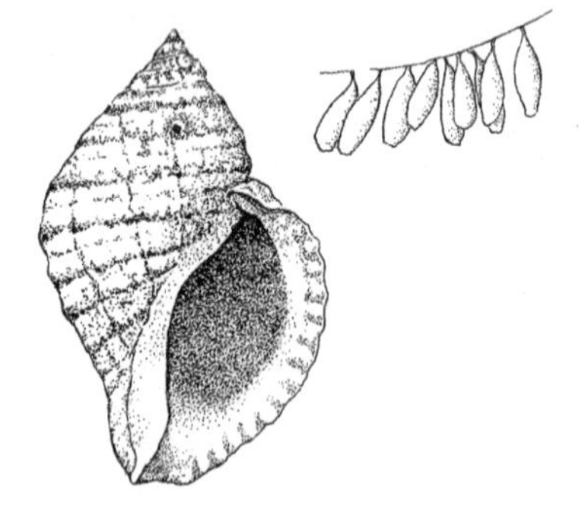

Dog whelk *(Nucella lapillus)*—up to 1.5 in (4 cm. Upper right, egg capsules attached under rock ledge.

a rock while rasping the surface with its teeth, known as a radula. Chitons usually eat single-celled or very small plants. They are represented locally by the **red chiton**, *Tonicella rubra*.

Gastropods are sometimes called univalves because they have one shell. Some snails, like nudibranchs, have lost their shells; in others, including bubble shells, the shell is reduced and covered with fleshy tissue. Whether they have a shell or not, all free-living snails (except the vermetids, or tube snails) have a flattened foot on which they glide over the bottom. They also have radulae, which can protrude from their mouths and be used to feed on everything—ranging from other animals, to plants and small microscopic particles.

The common predatory snails include the dog whelk, *Nucella lapillus*; the waved whelk, *Buccinum undatum*; the Northern moon snail, *Euspira heros*; the **ten-ridged whelk**, *Neptunea lyrata decemcostata*; Stimpson's spindle whelk, *Colus stimpsoni*; and the Atlantic oyster drill, *Urosalpinx cinerea*.

The **dog whelk** is encountered most frequently on the outer shore. It occurs in a variety of color patterns, ranging from black and white to orange. In some years, there are so many dog whelks that it seems the whole shore is covered with snails while, in other years, this species can be difficult to find. *N. lapillus* is the primary predator of barnacles and mussels on the rocky shore. Its young start their lives in egg capsules, which are carefully glued to the undersides of rock ledges in the intertidal zone.

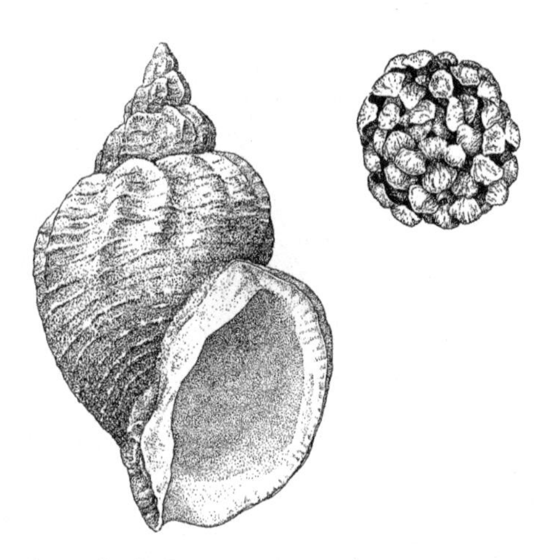

Waved whelk *(Buccinum undatum)*—up to 5.5 in (14 cm). Upper right, an egg case.

The **waved whelk** is a larger, heavier snail that is usually taken in subtidal samples, or from lobster traps where it feeds as a scavenger on lobster bait. It is less colorful than the other snails and is easily distinguished by the strong ridges running from the apex of the shell to the body opening. Egg cases of this species can be found on beaches after a late summer storm.

Stimpson's spindle whelk is found more rarely because it generally lives in slightly deeper waters. Still, it may be found occasionally washed up on beaches. Like other whelks, it is a carnivore.

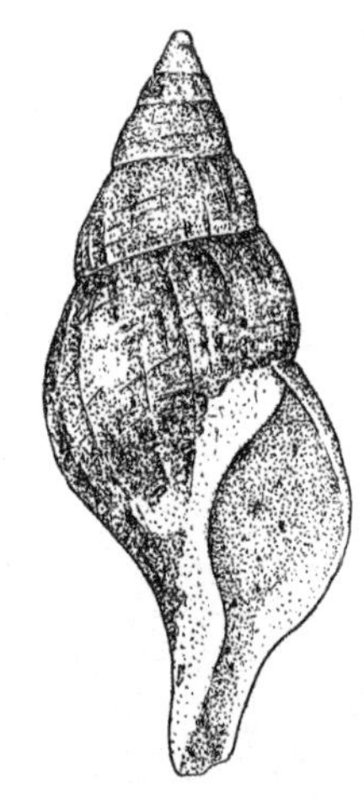

Stimpson's spindle whelk (*Colus stimpsoni*)—4 in (10 cm)

The smallest of the predatory whelks is the **oyster drill**, so-called because it uses its radula, in conjunction with a special enzyme-secreting gland near its foot, to

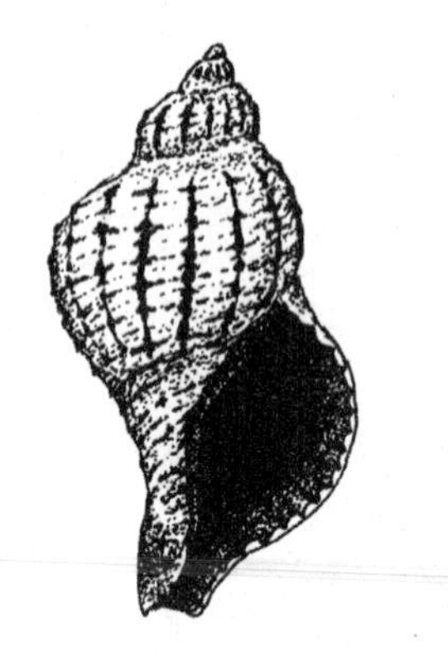

Atlantic oyster drill *(Urosalpinx cinerea)*—0.8 in (2 cm)

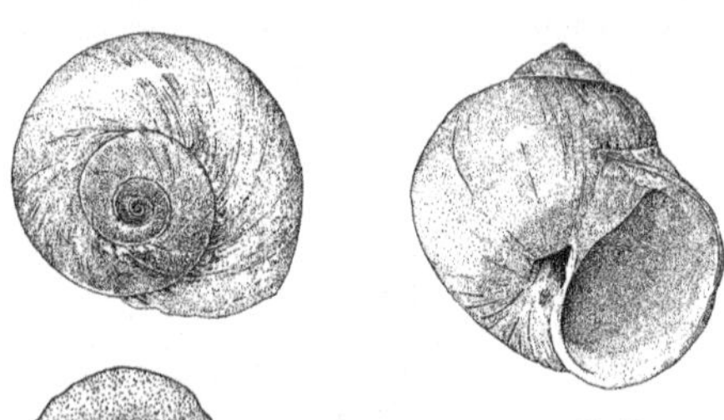

Northern moon snail *(Euspira heros)*—up to 4.5 in (11 cm). Left, sand collar with eggs.

Common periwinkle *(Littorina littorea)*—1 in (2.5 cm)

make a hole in oyster and other bivalve shells. In many areas of the U.S., **oyster drills** are serious economic pests. In Maine, they are common in areas where oysters were once abundant, but they also occur in the southern half of the state near aquaculture sites.

Moon snails are predators living on sandy flats or shallow, sandy mud bottoms where they feed primarily on bivalves. The northern moon snail, *Euspira heros*, may look more like an old piece of gristle that has been thrown overboard, as the body of the snail may obscure most of the shell. Like dog whelks, moon snails obtain their food by boring holes through their victim's shells and using their radulae to tear off pieces of meat until their prey is consumed. Moon snail eggs are laid in a collar of sand grains made by the female. These collars are often found on sand flats.

Many snail species are herbivores, feeding on the abundant algae of exposed and protected rocky shores. The most prominent herbivores are the littorines: *Littorina littorea*, *L. obtusata*, and *L. saxatilis*—the **common periwinkle**, smooth periwinkle, and rough periwinkle, respectively. *L. littorea* has a thick, smooth, purplish-brown shell and occupies the middle to lower levels of the intertidal zone. It is thought that *L. littorea* was introduced from Europe in the 1840s. However, there is new evidence suggesting *L. littorea* survived the last ice age in the Nova Scotia region and lived in relative obscurity until it was inadvertently spread by Europeans in the 1800s. It occurs in such large abundances that it can control easily the algal growth in areas where it occurs. The **smooth periwinkle**, *L. obtusata,* is a smaller species, orange to yellow in color, and it is usually found living on rockweed or knotted wrack on which it feeds. On the other hand, the **rough periwinkle**, *L. saxatilis,* is an inhabitant of the upper shore and usually occurs in protected areas (such as under rocks) or in the more protected reaches of coastal bays. Unlike the other two littorines, *L. saxatilis* produces young directly, skipping the normal free-living larval stages.

Smooth periwinkle *(Littorina obtusata)*—0.5 in (1.3 cm)

Rough periwinkle *(Littorina saxatilis)*—0.5 in (1.3 cm)

An herbivore of the lower shore, *Lacuna vincta,* looks like a littorine, but it is small and has a very thin shell that may be decorated with a delicate reddish-brown banding. Also, near the opening of the shell, there is a small slit-like hole (the umbilicus) that gives the snail its common name, **chink shell.** *L. vincta* is commonly found on kelp and

Northern lacuna or chink shell *(Lacuna vincta)*—0.5 in (1.3 cm). At left, egg mass on kelp. Lower right, actual size.

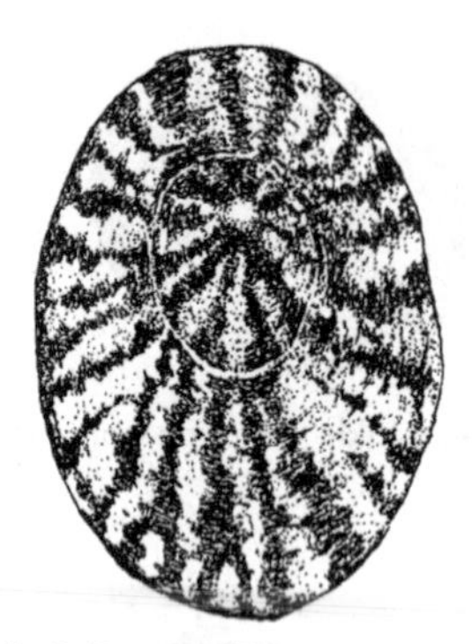

Tortoiseshell limpet *(Tectura testudinalis)*—up to 2 in (5 cm)

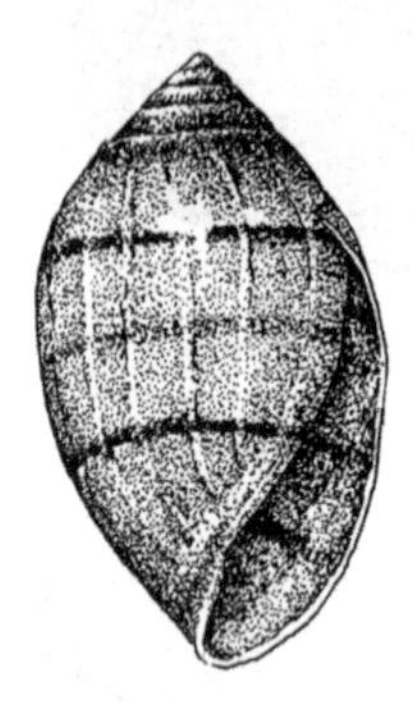

Leucophytia bidentatus—0.6 in (1.5 cm)

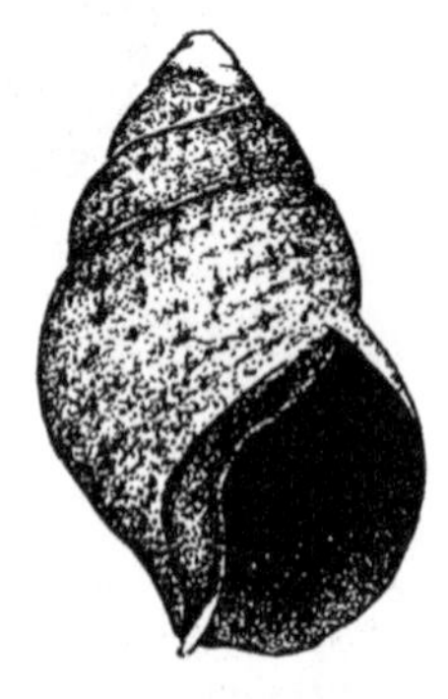

Eastern mud snail *(Ilyanassa obsoleta)*—1 in (2.5 cm)

sea grass on which it probably feeds.

Another common grazer of the rocky shore is the **tortoiseshell limpet**, *Tectura testudinalis.* Its shell has no signs of coiling, typical of snails, and the foot is very large. Limpets have especially strong teeth in their radulae, which they use to graze the coralline algae lining the bottom and sides of many rock pools in the lower intertidal zone. In fact, coralline algae need to have their outer layer of cells scraped off by limpets to be able to grow and reproduce.

Three snails of New England estuarine waters—*Leucophytia bidentatus, Ilyanassa obsoleta,* and *Hydrobia totteni*—specialize in grazing on small detrital particles, most likely gaining their nutrition from the resident bacteria, diatoms, and fungi. Animals that feed in this way are called microphages. ***Leucophytia bidentatus*** is a microphage with a stout, multi-banded shell that is most commonly found on the stems of salt-marsh grasses. In contrast, the **Eastern mud snail**, *Ilyanassa obsoleta,* has a dull, dark shell and is primarily found on muddy surfaces where it grazes on diatoms and other microalgae living among the sediment grains. In areas to the

pairs of antennae at the front of their heads, at least at some time during their lives.

While not obviously a crustacean, the common **barnacle** of the shore has all the typical crustacean features when it is in its larval stage. *Semibalanus balanoides* is the most abundant of all the local barnacles, occurring in a broad zone on the upper parts of rocky shores. The shell of this species is quite variable and, in fact, changes during the year. In the spring when it is newly settled, the shell looks like that of a typical acorn barnacle. As the summer progresses, the barnacles continue to grow and they gradually fill in all the space among themselves. They then have to grow upward. By early winter, all the individuals are very elongate and can be swept quite readily from rocks by waves or people walking on the shore. Barnacles are also unusual in that they have glued their "heads" to the rock and so have to feed with their "feet."

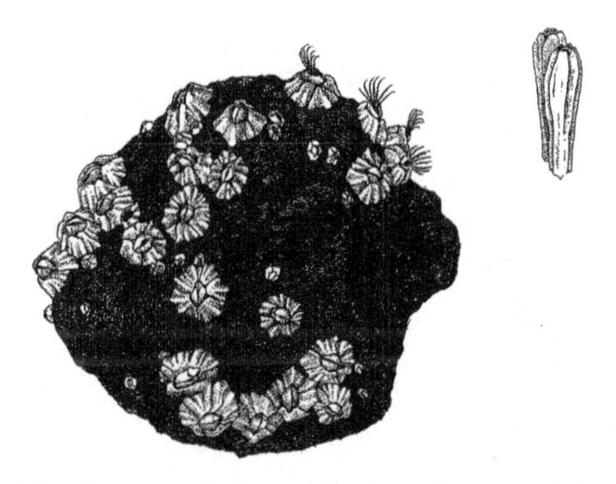

Northern rock barnacle *(Semibalanus balanoides)*—1 in (2.5 cm). Upper right, elongated form.

A major group of crustaceans are the **decapods**, so-called because they have 10 walking legs. In some cases, the legs may be modified to help capture food. Decapods are often grouped as either shrimp or crabs, with lobsters and hermit crabs falling between the two.

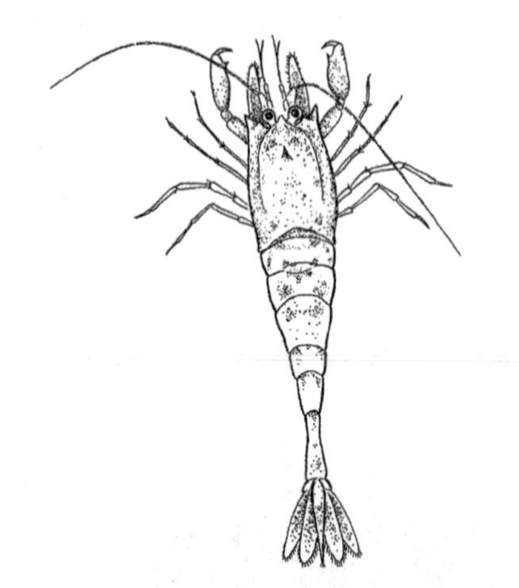

Sevenspine bay shrimp *(Crangon septemspinosum)*—2.8 in (7 cm)

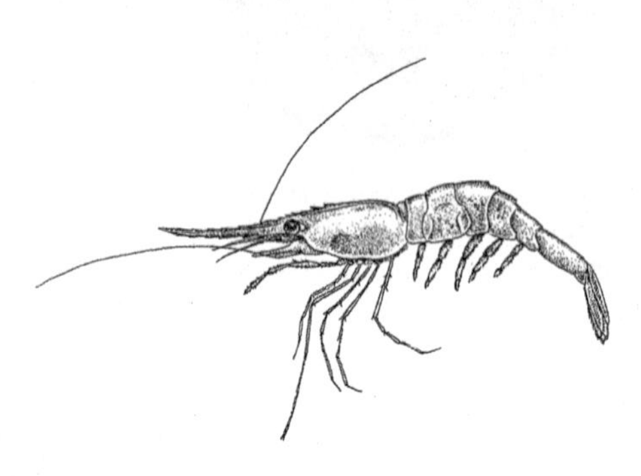

Northern shrimp *(Pandalus borealis)*—6.8 in (17 cm)

The most common estuarine shrimp is the **sevenspine bay shrimp**, *Crangon septemspinosum*. It is distinguished easily from all other shrimp by its front pair of legs that possess subchelae, or claws, that are flattened and have the moveable finger folded back against the base, or "hand." *C. septemspinosum* is also colored to blend into the background provided by the sand or mud on which it generally lives. Members of this shrimp genus are very abundant in estuaries and bays on both sides of the Atlantic, especially in Europe where *Crangon crangon* forms the basis of a commercial fishery.

The commercial **Northern shrimp** species in the Gulf of Maine is *Pandalus borealis*. It has a very long, saw-toothed rostrum (spike on the front of the head) and is usually some shade of red before being cooked. This species is widely distributed throughout the Arctic, North Atlantic, and North Pacific. In fact, the Gulf of Maine represents the most southerly occurrence of the species in the western Atlantic. As a result, most individuals are restricted to the deeper colder waters of the Gulf, until winter when the surface water cools. Then, the females move inshore where they release their larvae—if the shrimp have not already been removed from the sea by a trawler or trapper.

Of course, the largest and best-known crustacean in Maine is the **American lobster**, *Homarus americanus*. It has two large, pincer-like claws that draw much gastronomic attention. Lobsters, like shrimp, have elongated abdomens bearing swimmerets—small appendages under the abdomen (or tail) used for locomotion and which, in females, also carry eggs until the larvae hatch. The lobster's abdomen contains a large muscle that it uses to rapidly flex, or flap, the tail to escape predators or larger lobsters. In the case of human predators, the tail muscle is nearly useless as an escape device.

American lobster *(Homarus americanus)*—up to 34 in (86 cm)

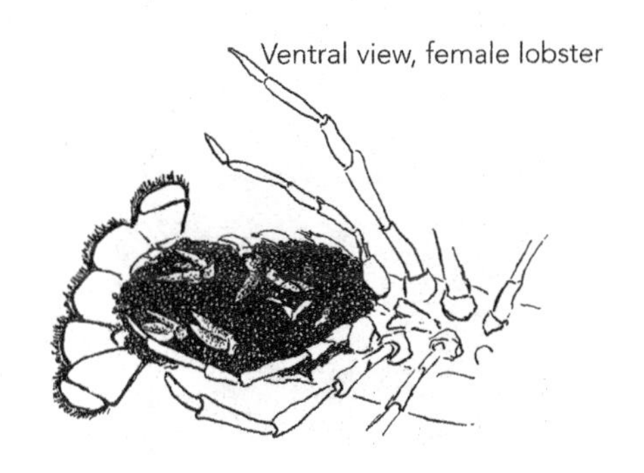

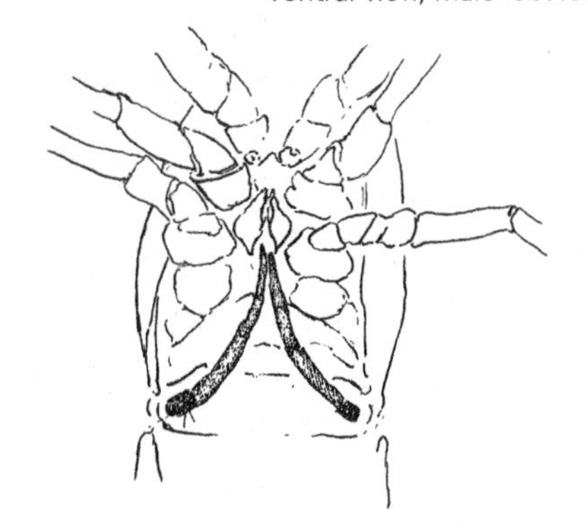

Hermit crabs are related more closely to lobsters than to the true crabs. All hermit crabs in Maine's colder waters inhabit gastropod shells, which they use to protect their soft abdomens. Five species of hermit crabs can be found in the Gulf of Maine, but two are particularly common. Details of their major and minor claws are often used to tell hermit crabs apart. *Pagurus acadianus* is a northern species whose claws are covered with low bumps. Its major claw has a prominent orange stripe and is about one-third larger than the minor claw. In

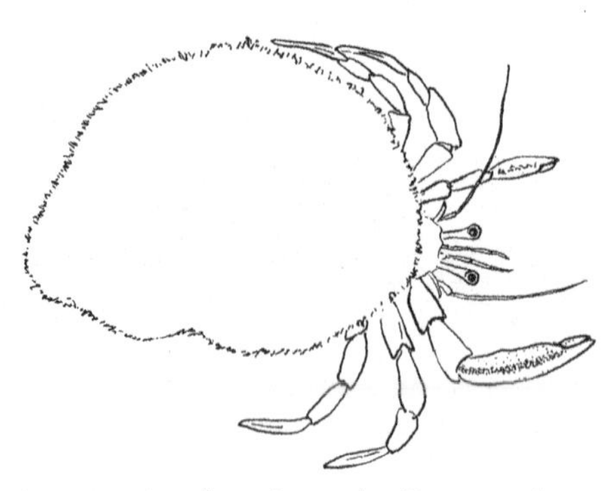

Long wrist hermit crab *(Pagurus longicarpus)*—0.3 in (0.8 cm)

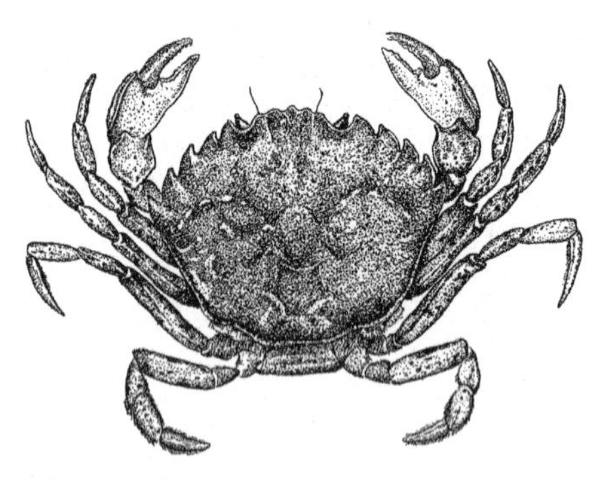

Green crab *(Carcinus maenas)*—up to 3 in (7.5 cm) carapace width. Below, ventral view of female. Left, male abdomen.

Pagurus longicarpus, on the other hand, the major claw is much longer than the minor, is nearly smooth, and has a brown or gray stripe. As noted earlier (see p. 9), snail shells inhabited by **hermit crabs** are often covered with a colonial hydroid in the genus *Hydractinia*. Each hermit crab species carries its own *Hydractinia* species with it.

A true crab has a box-like design, with its abdomen reduced in size and tucked under the front part of the body. The ten legs originate from the side of the body, which often gives crabs their odd sideways mode of walking. A crab breathes by using gills tucked safely under its carapace. Water pumped through the gill chamber enters at the base of the legs and exits in front of the mouth, producing the odd bubbling of water seen when crabs are kept alive in small amounts of water.

The most common crab of the shore is the **green crab**, *Carcinus maenas*. It is a member of the swimming crab family Portunidae; however, its last pair of legs is merely flattened instead of having fully developed paddles. As the name implies, green crabs are generally greenish in color, but they may also have tinges of yellow, especially on their undersides. Green crabs are invaders from

Europe, arriving most probably with early settlers. In southern New England and most recently in midcoast Maine, green crabs are being replaced in some habitats by another invader, *Hemigrapsus sanguineus* or Asian shore crab (not shown), which arrived a few years ago from Japan. The Asian shore crab can be distinguished by the box-like shape of its body and banded coloring of its legs.

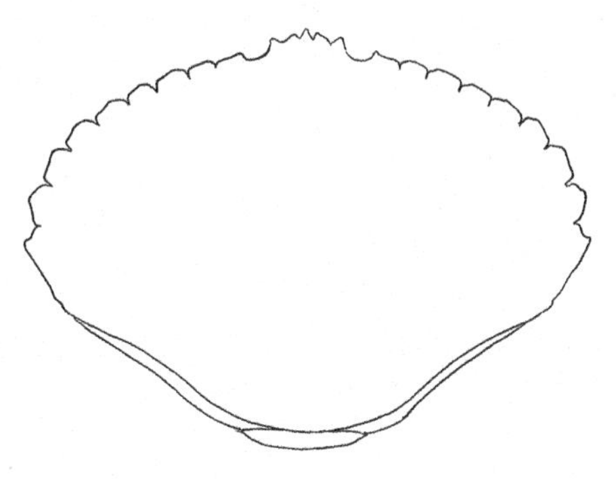

Atlantic rock crab *(Cancer irroratus)*—up to 5.5 in (14 cm) carapace width

The two commercially harvested crab species in Maine are rock and Jonah crabs, *Cancer irroratus* and *Cancer borealis*, respectively. The **rock crab** carapace is slightly yellowish with reddish flecks, and it is broadly oval with nine smooth or slightly granular teeth on each side of the front margin. The **Jonah crab**, on the other hand, is more reddish and the front margin teeth are rough-edged and jagged. If not caught in traps, rock and Jonah crabs may occasionally be found among algae on rocky shores. All individuals of both species molt within a few days of each other, often resulting in large numbers of crab shells being washed up on the beach.

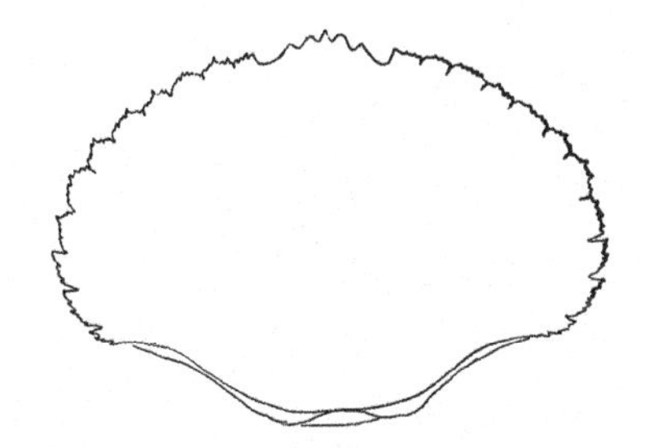

Jonah crab *(Cancer borealis)*—up to 6.3 in (16 cm) carapace width

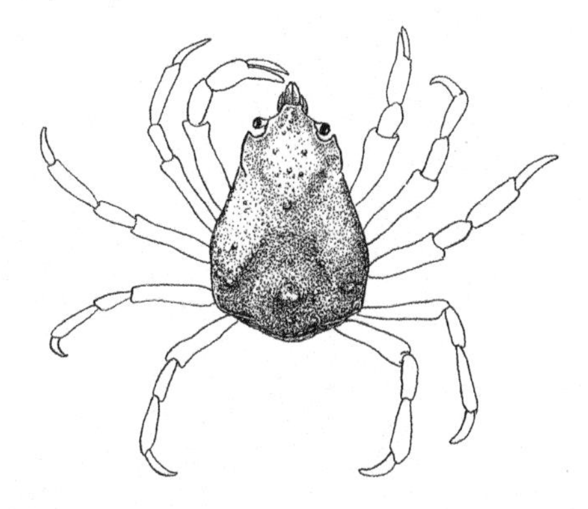

Arctic lyre crab *(Hyas coarctatus)*—1.3 in (3.3 cm) carapace width

The **Arctic lyre crab**, *Hyas coarctatus*, is a member of the **spider crab** family. It has a triangular carapace that is usually covered with a dense growth of algae, sponges, hydroids, or other species that can find a home among the hairs and spines on the crab's back. Spider crabs are slow-moving herbivores that use camouflage as their primary defense. As a result, their claws are relatively harmless to humans. *H. coarctatus* is usually found along the rocky shore, while its sister species, *H. araneus* (not shown), is usually found on softer substrates.

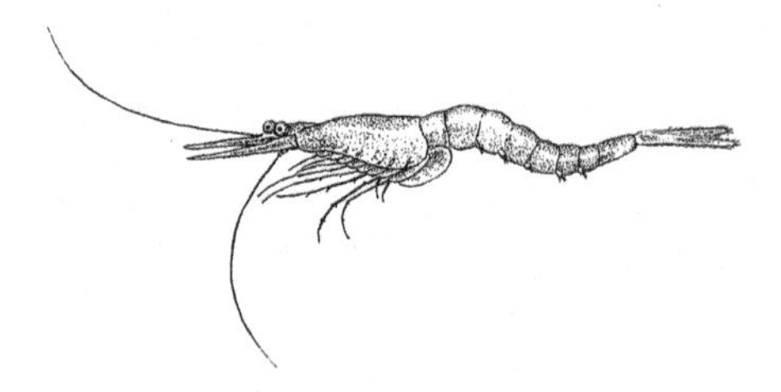

Bent opossum shrimp *(Praunus flexuosus)*—1 in (2.5 cm)

Mysids are shrimp-like crustaceans, but they differ from true shrimp in having seven pairs of walking legs and carapaces that are not attached to the backs of their bodies. Common inhabitants of shallow estuarine bottoms, mysids such as the **bent opossum shrimp** usually hover just over the sediment and feed on small animals and detritus. One of the largest mysids is the introduced European species, *Praunus flexuosus*. As with most mysids, it is nearly transparent when alive.

Amphipods are one of the most diverse groups of crustaceans on our shores. A typical amphipod has seven pairs of walking legs, arranged in three groups on the main part of its body, and its abdomen is divided into two sections. **Scuds**, members of the genus *Gammarus* are the amphipods most easily seen on algal covered shores. They are generally greenish in color and have a dark, kidney-shaped eye. Most *Gammarus* species are omnivores, eating plant material most of the time, but feeding on dead or dying animals when the opportunity arises. *Gammarus oceanicus* can be found in algae on the outer coast, *G. duebeni* in high intertidal pools on the outer shore, *G. finmarchicus* under stones or rockweed along protected shores and mudflats, and *G. tigrinus* under stones and algae in estuarine creeks. Another estuarine amphipod, common on tidal flats where the substrate is sandy mud, is ***Corophium volutator***. This amphipod makes a shallow (about 1 to 1.5 in or 2 to 4 cm deep) U-shaped tube, which it lives in when the tide is low. At high tide, it comes out of the tube and scrapes the mud surface into a small ball which it then takes into its tube and, using its front legs, sieves for edible organic matter, such as diatoms and algal pieces.

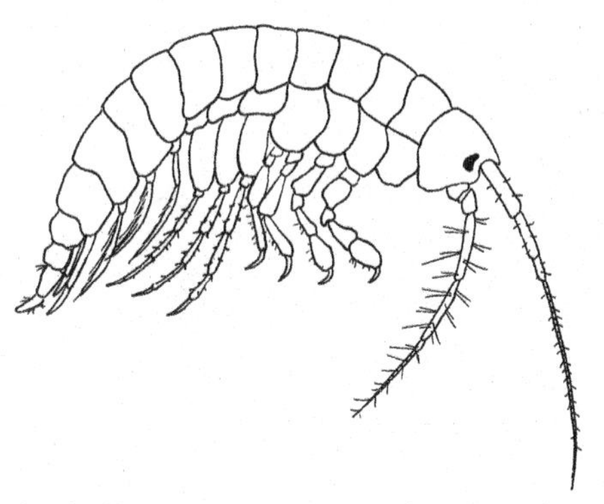

Scuds (*Gammarus* spp)—up to 1 in (2.5 cm)

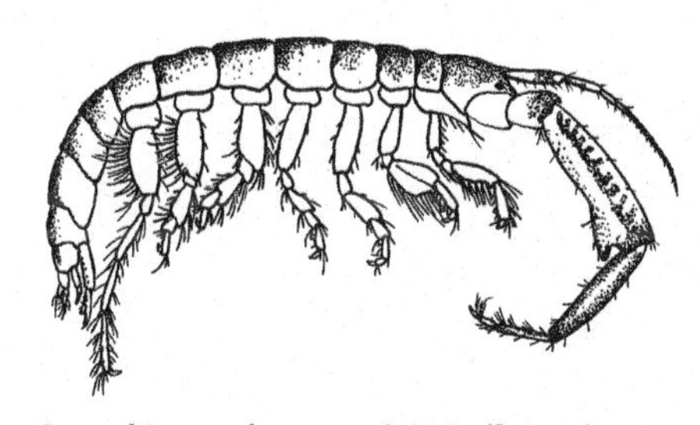

Corophium volutator—0.16 in (0.4 cm)

Another common group of amphipods includes the beachhoppers and allies in the families Talitridae and Hyalidae. ***Hyale nilssoni*** is a small, yellowish amphipod of the rocky shore, usually seen hop-

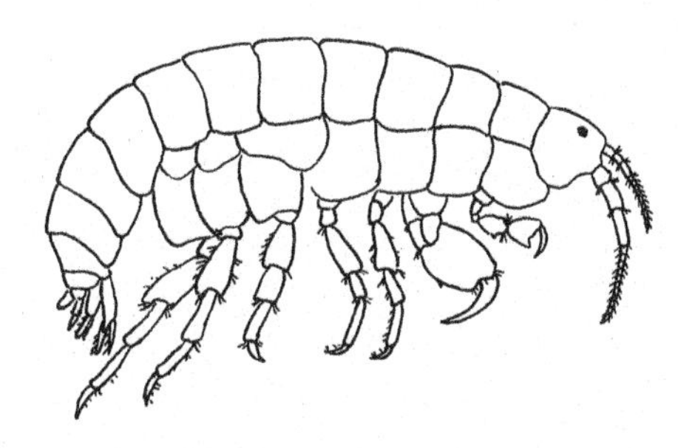

Hyale nilssoni—0.4 in (1 cm)

ping from one place to another like a large flea. ***Platorchestia platensis*** is a common inhabitant of the wrack line, especially where rockweed is washed up on shore. ***Orchestia grillus*** lives among the grasses and rotting vegetation of salt marshes. ***Ameriorchestia megalophthalma*** is a large (about 0.4 in or 1 cm long) beach hopper, best found by locating its oval-shaped burrow entrance on the higher areas of sandy shores. This species is also recognizable by its very large, round black eye.

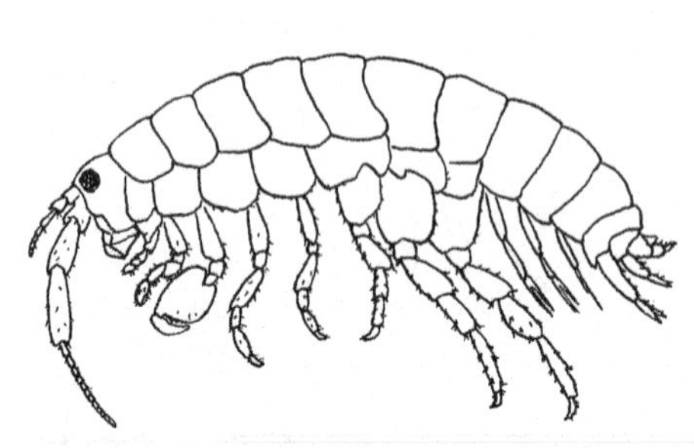

Platorchestia platensis—0.5 in (1.2 cm)

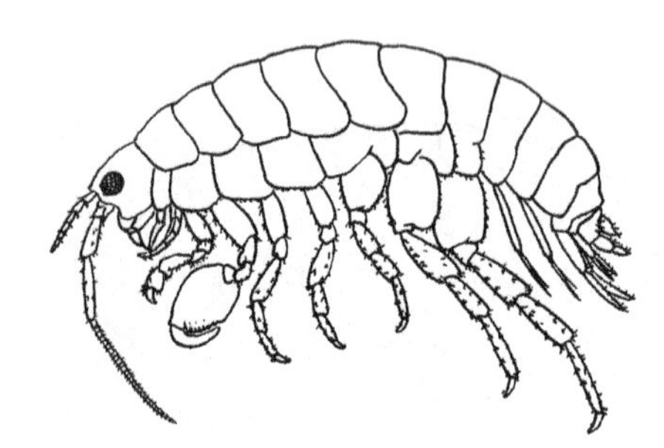

Orchestia grillus—0.8 in (2 cm)

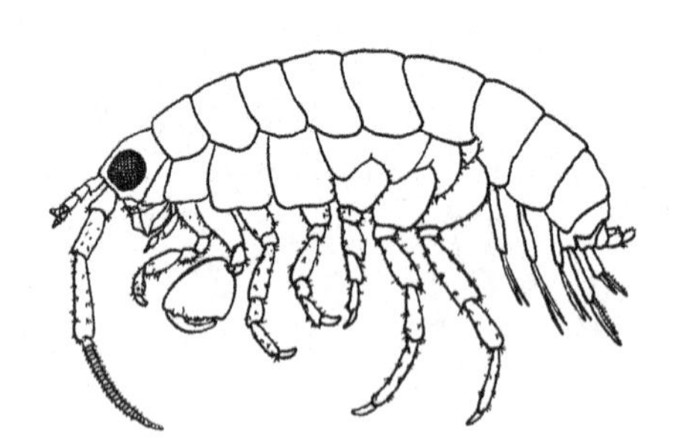

Ameriorchestia megalophthalma—1 in (2.5 cm)

Isopods, like amphipods, have seven pairs of walking legs but, in this case, the walking legs all look alike. Also, the isopod's abdomen is not subdivided, and its appendages are flattened and used as gills. There are two common species on the outer coast rocky shore: the larger, reddish-purple ***Idotea baltica*** and the very small, ovoid, and dark gray ***Jaera marina.*** *I. baltica* is always found living in algae, whereas one can see *J. marina* most easily by looking on the undersides of rocks immersed in tidal pools. The little black specks seen moving around on the rock surface are *J. marina.* In estuaries, *I. baltica* is also found living on eelgrass, in which case it is usually greenish in color.

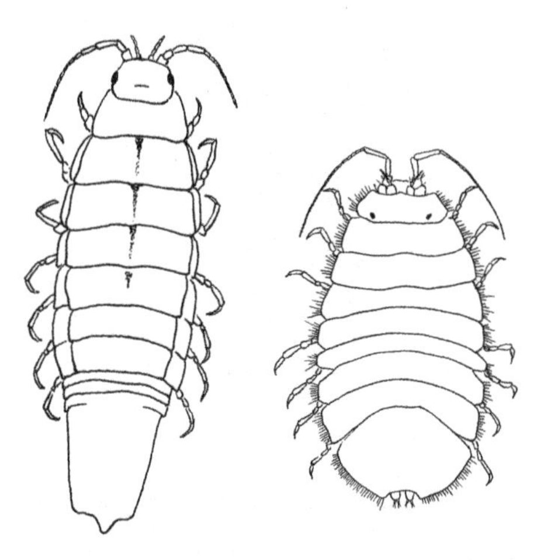

Left, ***Idotea baltica***—1 in (2.5 cm), and right, ***Jaera marina***—0.16 in (0.4 cm)

Two other isopod species can be found high on the shores in estuarine areas, but careful sleuthing is required to find them. In some books, these species are considered to be terrestrial in their habits, but both can be found commonly in algal wrack or under washed-up pieces of wood. *Ligia oceanica*, the **rock slater**, is the more difficult species to find. It generally hides during the day in cracks in open rock ledge about 3 ft (1 meter) or so above the high water mark. If you gently move smaller rock pieces in these cracks, one or two individuals

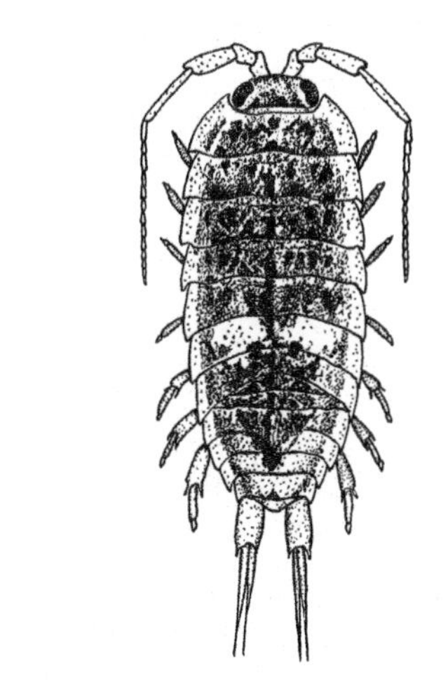

Rock slater *(Ligia oceanica)*—1 in (2.5 cm)

Common rough wood louse *(Porcellio scaber)*—0.4 in (1 cm)

usually will be rousted from their hiding places. The **rough wood louse**, *Porcellio scaber*, is a common summer dweller of moist algal wrack, but it can also be found several meters away from the water, under leaves and logs in the adjacent forest. This species lives on decaying vegetation and requires some moisture for its survival.

Echinoderms are truly unusual invertebrates. They have a five-part symmetry to their bodies and do not have a brain comparable to that seen in worms, mollusks, or crustaceans. Most echinoderms also have their bodies completely covered in small calcareous plates through which tube feet, used for locomotion, protrude. There are four common groups of echinoderms that can be found in Maine waters: sea stars, brittle stars, sea cucumbers, and sea urchins and sand dollars.

Sea stars are typical echinoderms. They usually, but not always, have five arms, and each arm usually has a complex arrangement of tube feet lining its undersides. The upper side of a sea star's body is covered with a dense set of calcareous plates, set so closely together that it appears the animal has a continuous coat of armor. Also on the upper surface is a small button-like structure, called the madreporite, which the animal uses to take in water; this helps the tube feet move. *Asterias rubens,* the **Northern sea star,** has a pale yellow madreporite, whereas the one on *Asterias forbesi,* the **common sea star,** is bright orange. Otherwise, the two species are very difficult to tell apart.

Five-Part Harmony

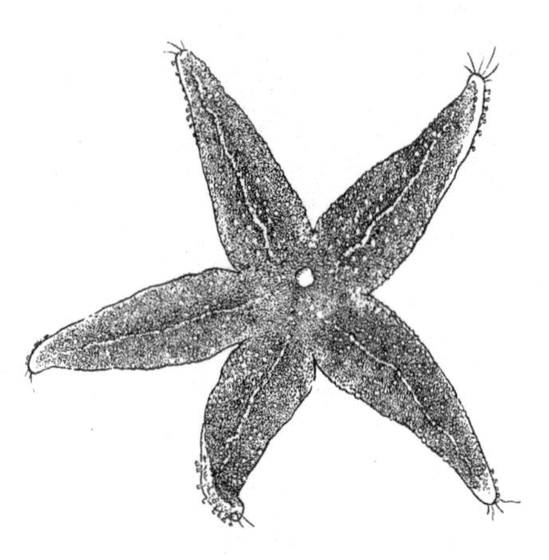

Northern sea star *(Asterias rubens)*—up to 8 in (20 cm)

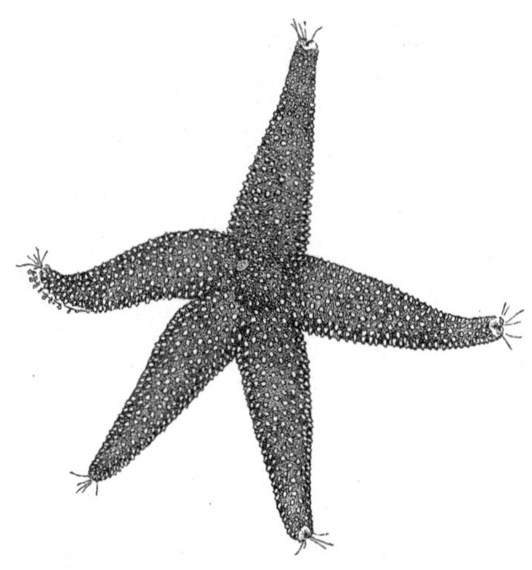

Common sea star *(Asterias forbesi)*—up to 5 in (13 cm)

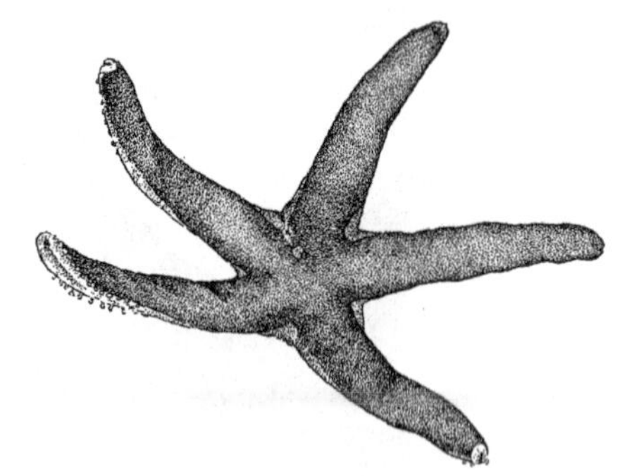

Henricia spp.—up to 4 in (10 cm)

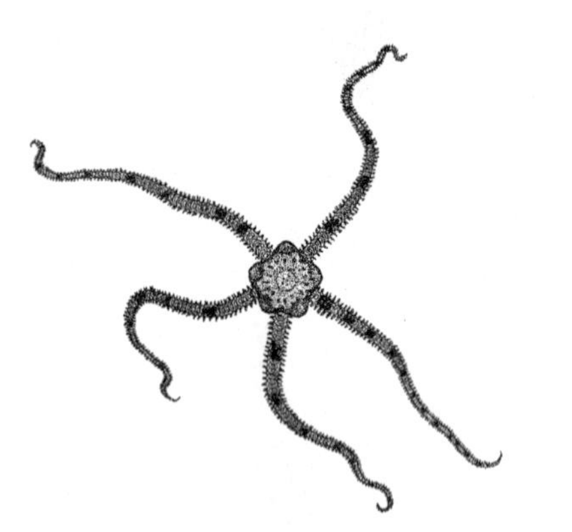

Daisy brittle star *(Ophiopholis aculeata)*—0.8 in (2 cm) across disk

Henricia spp., on the other hand, have only two rows of tube feet along the underside of the arms, and occasionally one of these animals may be found with six arms. The upper part of the body in this group of sea stars is also generally quite smooth.

Brittle stars are built completely differently from true sea stars. Their arms and bodies are distinct from each other, and the calcareous plates lining the arms have several different designs. Brittle stars get their name from the fact that their arms generally break away from their bodies relatively easily. The most accessible local brittle star is the daisy brittle star, *Ophiopholis aculeata*, which can be found under rocks and sponges low in the intertidal zone.

Sea urchins and sand dollars, unlike all other echinoderms, have bodies that apparently are completely encased in calcareous plates attached to each other without flexible joints. While it appears that the hard plates are on the "outside" of the animals' bodies, there is actually a thin layer of skin covering the plates. Both species have calcareous spines attached by small muscles to the hard plates, but those of the **sand dollar**, *Echinarachnius parma*, are shorter than those of the sea urchin, *Strongylocentrotus droebachiensis*. Sand dollars, as the name implies, live on open sandy bottoms of bays and offshore banks and feed on small plant particles that they catch from the water passing over their bodies.

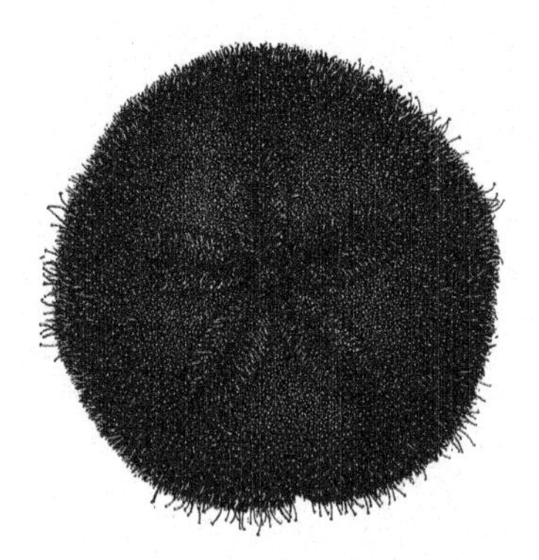

Sand dollar *(Echinarachnius parma)*—up to 3 in (7.5 cm) alive *above*; below, sand dollar skeleton.

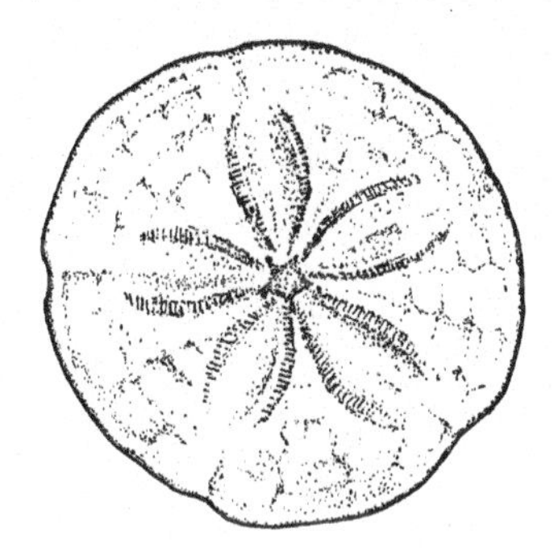

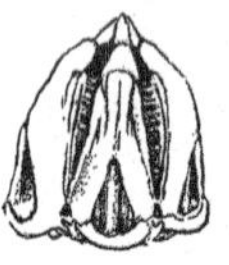

Green sea urchin *(Strongylocentrotus droebachiensis)*—up to 3.5 in (9 cm); alive above; left, Aristotle's lantern; below, sea urchin skeleton.

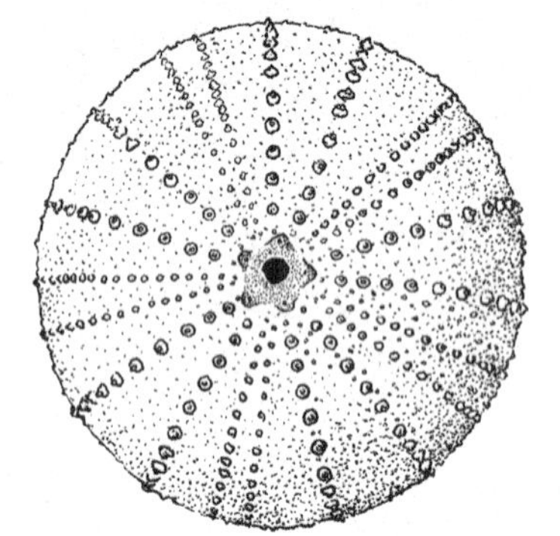

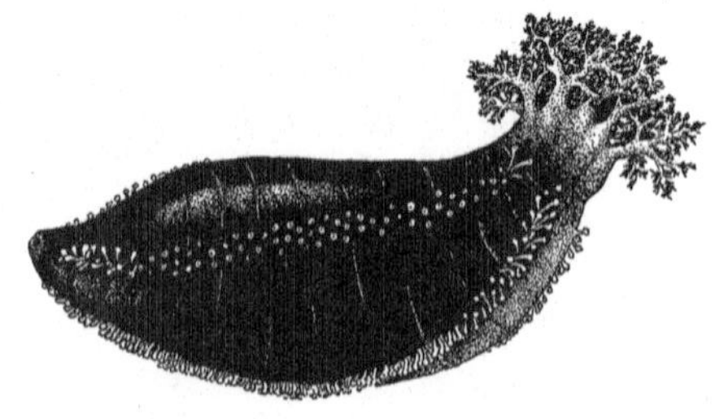

Orange-footed cucumber *(Cucumaria frondosa)*—up to 19 in (48 cm)

Sea urchins live near rocky surfaces where they feed on larger algae, rasping away at the algal surface with a series of strong teeth anchored in a large structure around the mouth called "Aristotle's lantern."

Sea cucumbers, represented in New England by the common species, *Cucumaria frondosa*, have long cylindrical bodies with tube feet arranged in rows running the lengths of their bodies. At the anterior end of its body is a set of tentacles that the sea cucumber uses to obtain its food. When undisturbed, *C. frondosa* extends its tentacles into the flow of water and waits for small food particles to stick to their mucus coating. Every once in a while, the sea cucumber bends a tentacle toward, and into, its mouth and "licks" off the adhered particles. Another peculiar feature of sea cucumbers is that the entrance to the respiratory system and the exit to the gut are both located within the same aperture at the posterior end of the animal. Sea cucumbers respire by pumping water into a delicate structure, called the respiratory tree, which exits next to the anus.

Chapter Three

COMMON FISHES OF TIDE POOLS AND SALT MARSHES

John Moring, University of Maine

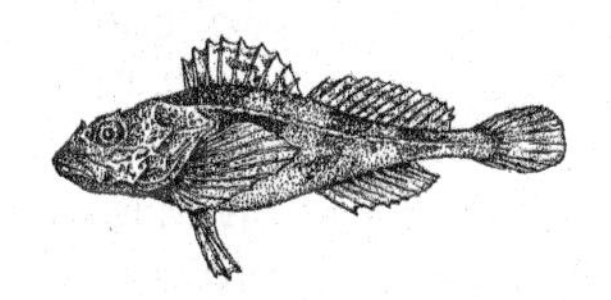

The intertidal zone is a very dynamic and productive part of the ocean. It is also the part of the sea that is most accessible to humans. The most common fishes along the coast of Maine are typically found in estuaries—where freshwater streams enter the ocean—as well as in bays and intertidal zones. Fishes that are found in tide pools, trapped bodies of water left in the intertidal zone at low tide, are the most accessible. Some fish species are found in tide pools most of their lives. Others use tide pools as refuges during juvenile stages or move into the intertidal zone only with flood, or high, tides. In addition, many types of inshore fishes forage in the intertidal zone and return to subtidal waters during low tide. Following are descriptions of fish species that are especially common in tide pools, salt marshes, bays, and estuaries.

Fishes

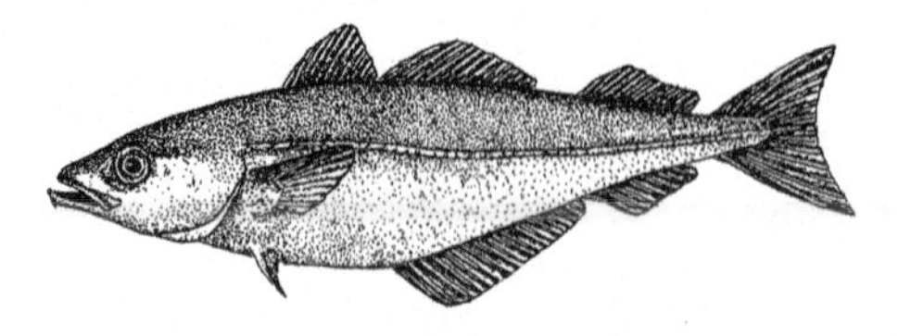

Pollock *(Pollachius virens)*—adults up to 3.5 ft (1.1 meter) in deeper waters, but generally much smaller; juveniles usually under 8 in (20 cm) inshore

The **pollock**, *Pollachius virens*, is extremely abundant in shallow waters almost everywhere along the coast. Sometimes called the harbor pollock, this fish is one of the most common fishes caught by coastal anglers. Juvenile pollock can be seen swimming in schools near docks and piers, and they often invade tide pools where they seek refuge in clumps of rockweed. As pollock increase in size, they adapt to a more benthic life and move into deeper waters. Adults live in waters as deep as 480 ft (146 meters) and are commercially important. However, the young fish typically are taken only by sport anglers. These juveniles have a distinctive copper-brown coloration, with silvery-white undersides.

The pollock is a member of the cod family and resembles haddock and cod in shape. It has a distinct lateral line, similar to a dotted line on each side of its body, which contains sensory receptors. Like cod and haddock, pollock have three dorsal fins and two anal fins.

There is a small barbel (like a chin whisker), but it may be difficult to see in small fishes and often is absent in large fishes. Pollock have noticeably forked tails.

The **winter flounder**, *Pleuronectes americanus*, is the most common flatfish in inshore waters. With a distinctive oval or diamond-shaped body, the winter flounder has both eyes on the right side of its head. At about five to seven weeks of age, the winter flounder turns on its side and undergoes a metamorphosis in which the skull twists and the left eye moves alongside the right eye. The fish actually swims on its side and rests on its left side while on the bottom. The blind side (left) is opaque or cream-colored, while the upper side (right) has a darkened appearance that allows the fish to blend in with the ocean bottom, camouflaging it from predators. Although winter flounder are found in waters as deep as 120 ft (37 meters) inshore, the largest specimens are encountered on Georges Bank. Winter flounder move into the intertidal zone to feed during high tides, and juveniles live in shallow, sand-covered tide pools, as well as in inshore sandy areas.

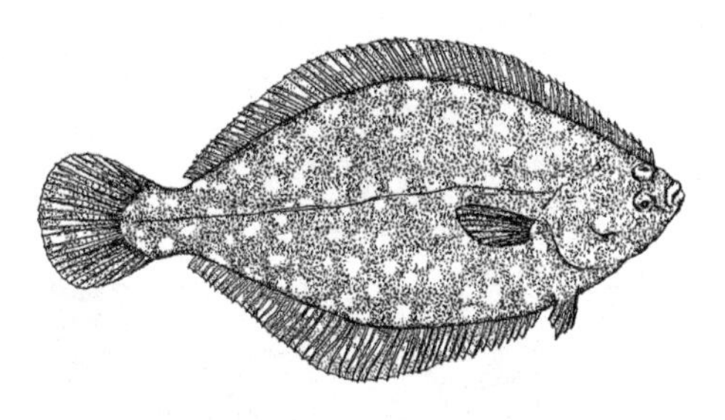

Winter flounder *(Pleuronectes americanus)*—adults up to 18 in (46 cm) inshore; juveniles typically less than 2.7 in (7 cm) in tide pools

The lateral line is straight in winter flounder and the area between the eyes has scales.

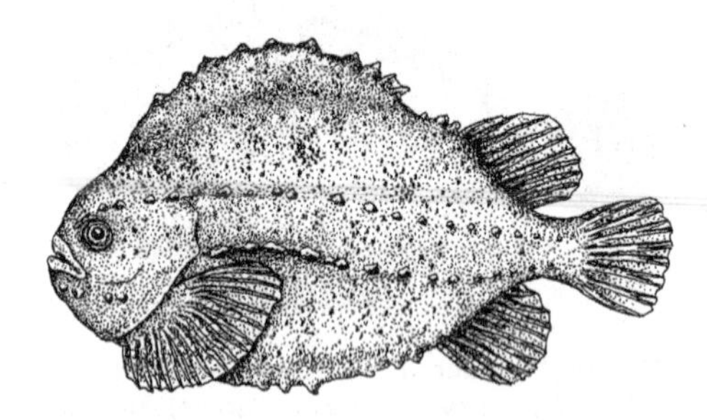

Lumpfish *(Cyclopterus lumpus)*—adults up to 23 in (58 cm); juveniles in tide pools 0.3 to 2.7 in (0.7 to 7 cm)

Resembling a "football with warts," the **lumpfish**, *Cyclopterus lumpus*, is a most unusual-looking fish. Juvenile lumpfish are often found attached to seaweed, either in tide pools of the rocky intertidal zone or floating in the open ocean. After the first summer or two in tide pools, lumpfish move into subtidal waters where they are sometimes caught by anglers or found attached to lobster traps. Lumpfish spawn near shore or in tide pools in the spring, and the males guard the adhesive egg masses until the eggs hatch. In turbulent inshore waters, a lumpfish maintains its position by using a sucker structure on its ventral (or bottom) surface. Young lumpfish use their suckers to attach to specific types of marine algae, and the adhesion is remarkably powerful. Adult lumpfish are sometimes caught by anglers fishing along the rocky shore.

The lumpfish has a series of tubercles, or rows of bumps, along the length of its body. Such tubercles are not always developed in fish less than one year old. The lumpfish's most unique feature is a suction device on the underbelly. Only the Atlantic seasnail and other members of the snailfish family have this feature. The body color of lumpfish is variable, depending on the color of the surrounding habitat and the sex of the fish; it can range from green or gray, to brown or red.

Like the lumpfish, the **Atlantic seasnail**, *Liparis atlanticus*, possesses a ventral suction device, which it uses to attach to marine algae and other kinds of habitat. But a seasnail looks quite different from a lumpfish. The Atlantic seasnail has no rows of tubercles, and its body is more elongated with fragile-looking, almost jelly-like flesh. Other snailfish species are found throughout the world, but the Atlantic seasnail is the most common inshore species found in Maine waters. It seldom appears in tide pools before August, but it can be one of the more common species found there in late summer and fall. Because it hides in amongst seaweeds of various colors, it can adopt different color variations, which protects it from predators.

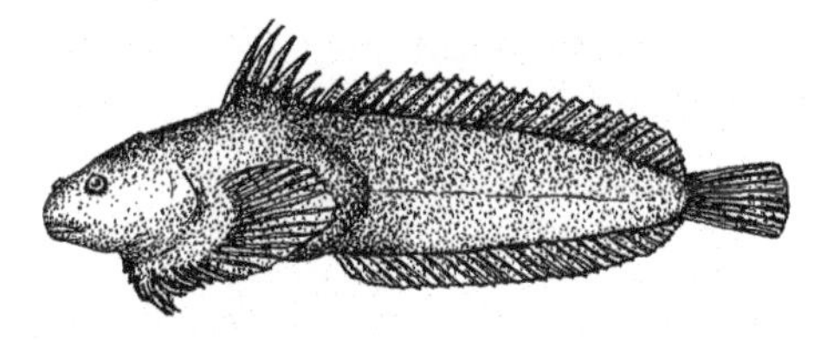

Atlantic seasnail *(Liparis atlanticus)*—adults up to 5 in (12.7 cm); juveniles as small as 0.4 in (0.9 cm) in tide pools

Seasnails typically are found in tide pools, attached to species of brown algae. The body shape is streamlined, except for the suction structure on the underside. The seasnail's body color ranges from dark or light brown to red, and it has a notched dorsal fin.

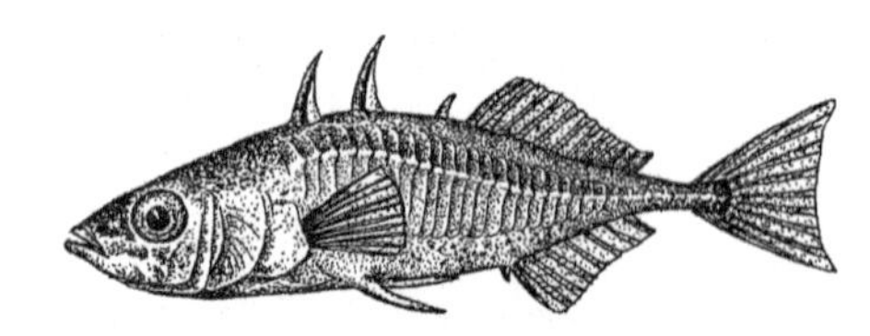

Threespine stickleback *(Gasterosteus aculeatus)*—up to 4 in (10 cm)

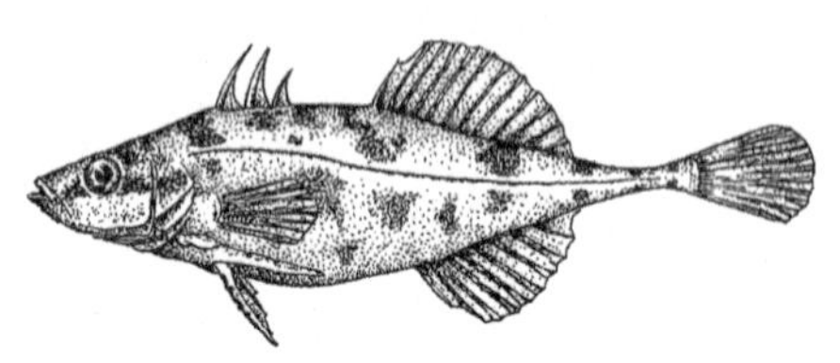

Fourspine stickleback *(Apeltes quadracus)*—up to 2.5 in (6.4 cm)

There are four species of **sticklebacks** found in Maine waters, but only three species are common in inshore waters. The threespine stickleback, *Gasterosteus aculeatus*, and the fourspine stickleback, *Apeltes quadracus*, are found in freshwater, brackish, and inshore marine waters, and they can be very abundant in salt marshes. The ninespine stickleback, *Pungitius pungitius occidentalis* (not shown), is abundant in estuaries and marshes, but it is rarely found in tide pools. Scientists have studied extensively the stickleback's unusual nest-building and spawning behavior. Sticklebacks are prey to many marine predator fishes. These fishes are characterized by their dart-shaped bodies; plate-like structures on their sides; and obvious, individual spines on their backs. The ninespine stickleback has dorsal spines set in a zigzag line, leaning alternately to each side.

Threespine sticklebacks are quite variable in color and prefer weedy areas and clumps of floating seaweed. Fourspine sticklebacks also are found near vegetation, but they depend less on algae than do threespine sticklebacks. Threespine sticklebacks are quite abundant in freshwater, and less so in brackish waters; while **fourspine sticklebacks** are typically found in brackish water, with occasional forays into freshwater. Ninespines are encountered in marine, brackish, and fresh water. During breeding times, hundreds of sticklebacks may invade shallow waters and tide pools.

Sticklebacks are characterized by very narrow caudal peduncles (the area of the body just in front of the tail). The threespine stickleback has three, very obvious, free spines on the top; the fourspine stickleback has four free spines; and the ninespine has nine or more free spines.

The **mummichog**, *Fundulus heteroclitus*, is extremely common along the coast, found mainly in shallow waters of tide pools and salt marshes. It is related to the banded killifish that is found in fresh water and in brackish areas of estuaries, and resembles a guppy or mosquitofish in appearance. An extremely adaptable fish, the mummichog can withstand wide variations in environmental conditions. It is often the dominant fish species in salt marshes and tidal creeks and moves back and forth with the tides.

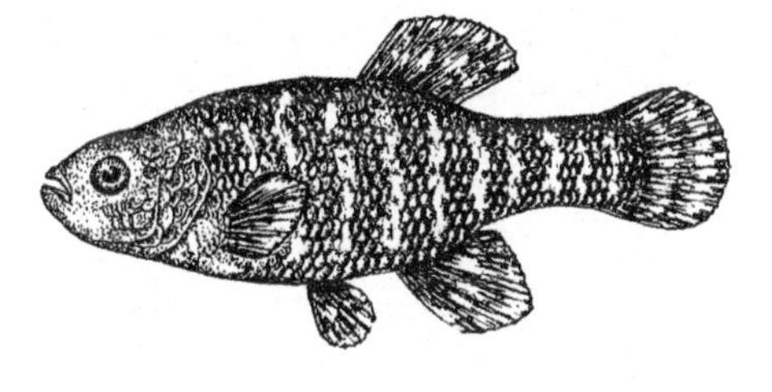

Mummichog *(Fundulus heteroclitus)*—usually up to 4 in (10 cm)

Large, distinctive scales on the mummichog's side give the appearance of cross-hatching. This fish has a blunt head, a dark or green color on its back, and silver bars on the side of its body. A mummichog has a thick caudal peduncle, the area of the body just in front of the tail.

There are several species of **sculpins** found in coastal waters, but the **grubby**, *Myoxocephalus aenaeus*, and the **shorthorn sculpin**, *Myoxocephalus scorpius*, are important members of tide pool fish communities. Both species have antifreeze components in their blood plasma that allows them to live in shallow waters where water temperatures can be near freezing. Almost every moderate-sized tide pool in the rocky intertidal zone contains one or more sculpins of these two species. It is sometimes difficult to distinguish grubbies from shorthorn sculpins when they are small, except by counting their anal fin rays.

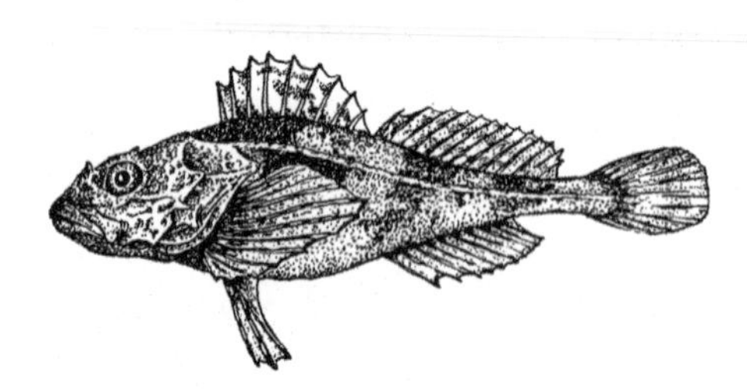

Grubby *(Myoxocephalus aenaeus)*—generally 2 in (5 cm) in tide pools

Grubbies never reach large sizes, but they can become quite abundant in bays, estuaries, and tide pools. The **shorthorn sculpin** is more associated with rocky coasts, can be as large as 36 in (90 cm), and is found in waters as deep as 360 ft (110 meters). Shorthorn sculpins over 3.9 in (10 cm) in length are sometimes found in tide pools, where it is an advantage to be small. Both sculpin species rest on the bottom of tide pools or hide amongst clumps of marine algae. Their mottled black-and-white coloration often camouflages them against the bottom. Grubbies have been found down to depths of 420 ft (130 meters), but small sculpins are typically found in tide pools and estuaries.

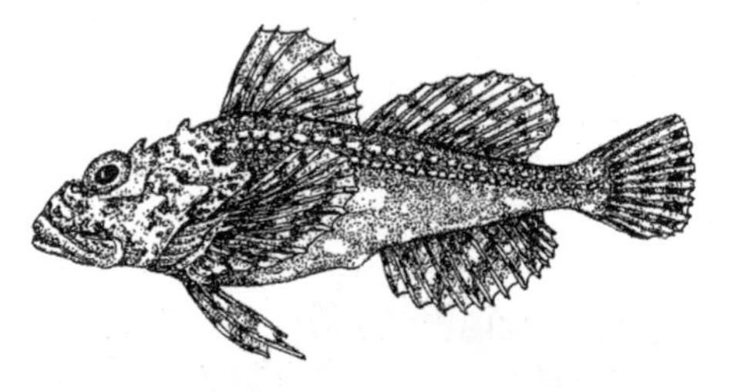

Shorthorn sculpin *(Myoxocephalus scorpius)*—typically smaller than 6 in (15 cm) inshore

Both grubbies and shorthorn sculpins can adapt to the turbulence and variable salinities and temperatures of the coastal environment, and both are noted for their broad diets. They can live in many environments and eat almost anything that is available.

The grubby has 10 to 11 anal fin rays, while the shorthorn sculpin has 13 to 14 anal fin rays. Most sculpins larger than 2.8 in (7 cm) found in tide pools or salt marsh pools are likely to be shorthorn sculpins. The grubby is more compact and tadpole-shaped than is the shorthorn sculpin.

The **Atlantic silverside**, *Menidia menidia*, is commonly found in salt marshes and estuaries where it is one of the most important prey species for inshore marine fishes. A schooling fish, it is observed often along sandy beaches and gravel-bottomed stretches of salt marshes, but it is found also in brackish waters and river mouths. Silversides prefer

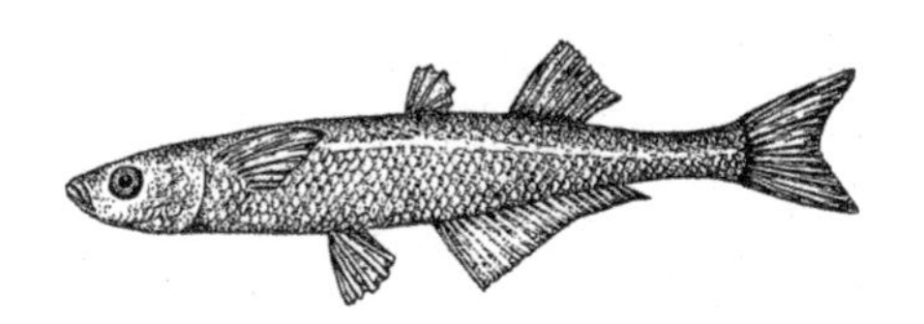

Atlantic silverside *(Menidia menidia)*—adults up to 5.5 in (14 cm)

shallow water (less than 6.5 ft or 2 meters deep) and have a varied diet. A silverside has a dark-colored back and a silvery stripe on its side. It has a rounded belly and two separate dorsal fins. The first dorsal fin is quite small and often difficult to see in small specimens.

Rock gunnel *(Pholis gunnellus)*—typically less than 6 in (15 cm) in tide pools or under intertidal rocks

Although the **rock gunnel**, *Pholis gunnellus*, looks like an eel, it has most of the fins found on other fishes. With its long, snake-like shape, the gunnel is most commonly found under rocks in the intertidal zone. Gunnels live in many parts of the world, but the rock gunnel is the only member of the family in the Maine intertidal zone. The same species is found along the coast of Great Britain, where it spawns intertidally in winter. Water temperatures are too low along the Maine coast in the winter, and rock gunnels move just offshore. However, they return to the same areas and even the same rocks in the spring.

Scientists are particularly fascinated with rock gunnels because they are able to live nearly out of water for hours during low tides, as long as they are protected in damp areas beneath rocks. Gunnels also are found in amongst clumps of marine algae in tide pools where they feed on small crustaceans.

Rock gunnels are one of very few eel-shaped fishes that are found along the coast. However, unlike an American eel, a rock gunnel has a long dorsal fin with spiny fin rays and a small pair of pelvic fins. A rock gunnel has a distinctive series of half-moon-shaped marks along its dorsal surface. Its coloration is usually dark, but it can vary between shades of red and green, depending on the algal background where the gunnel is hiding. There is also a distinctive dark bar that extends downward from the eye. Rock gunnels are sometimes known as "butter fish" in England because they have a slime coating that makes it very difficult for a human to hold a gunnel in one hand.

The **American sand lance**, *Ammodytes americanus*, is a long, thin fish that is seen most often darting across sand-covered pools and burying itself in the sand. Occasionally, sand lances are seen burrowing into the sand in the surf zone along a beach, where they can disappear beneath several inches of sand within seconds. This burrowing behavior is perhaps their most unique trait. Sand lances are quite common in salt marsh areas with sand bottoms and are important food items for larger fishes and even whales.

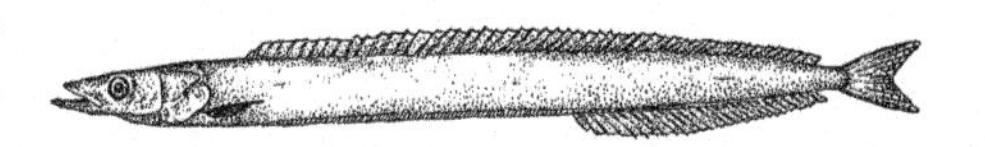

American sand lance *(Ammodytes americanus)*—typically less than 6 in (15 cm)

Sand lances are primarily found in full-strength seawater and are found less commonly in brackish water. They have a long, thin shape; dorsal fins without spines; and no pelvic fins. In sand lances, the lower jaw extends past the upper jaw and there is a single, straight lateral line high along the body.

Other Fishes

Inshore anglers sometimes catch **cunner**, a common subtidal fish, and **Atlantic herring** are seen often in tide pools and salt marshes. The **sea raven**, another type of sculpin, is very common in the intertidal zone where it forages at high tide. Rarely, sea ravens become trapped in tide pools, but this ferocious-looking fish (a body full of spines and bumps and a mouth full of teeth) is simply too large to remain hidden at low tides. Sea ravens are quite docile, however, and inshore anglers sometimes catch them on rod and reel. **Rainbow smelt** are found in estuaries and inshore waters and, like their landlocked cousins, return to rivers to spawn in the spring. The **American eel** is sometimes encountered in tide pools and inshore waters as it moves from the ocean to freshwater (elver stage) or in its larger, immature stages. The **Atlantic tomcod**, a relative of the pollock, is relatively common in river mouths and estuaries, and it will travel into freshwater portions of rivers.

Chapter Four

VASCULAR MARINE PLANTS AND SEAWEEDS

Jill Fegley, Maine Maritime Academy

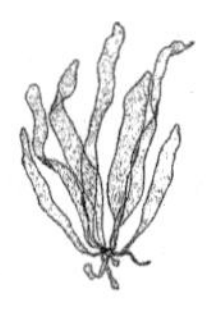

VASCULAR MARINE PLANTS

Very few species of higher plants have adapted to the marine habitat. Those that can live in, or near, salt water have special forms and structures that allow them to cope with both salt stress and low oxygen levels in the sediment. The extent to which different species can adapt to these stresses determines where they live. Salt marsh plants are periodically inundated by the tide, whereas sea grasses are completely submerged in seawater. Both types of marine plants are found along the New England coast.

Salt marsh communities are commonly found along the margins of temperate estuaries, especially in sheltered areas. Salt marshes form where there is little disturbance from water motion and ice, which allows sediments to accumulate and vascular, or flowering, plants to grow. In northern climates such as New Hampshire, Maine, and Canada, the scouring action of winter ice erodes away the seaward borders of developing marshes annually, leaving them in a constant state of recovery.

Tidal marshes are defined by the kinds of plants that can grow there. Drainage, salinity, and topography are particularly important in the distribution of salt marsh plants. Salt marshes basically can be divided into three fundamental zones: low marsh, high marsh and upland. Low marshes are younger, occupy a topographically lower position, and are more marine (containing salt water) or estuarine (having a mixture of fresh and salt water) than high marshes. Low marshes are also usually flooded at least once each day. The boundary between high and low marsh occurs at about mean high-water neap tide. Neap tides are those that have a minimum range between low and high tides, and they occur at the first and third quarters of the moon. High marshes are older, occupy a higher topographical position, and are influenced more by terrestrial or land conditions than are low marshes. They are only submerged during high spring tides (those having a maximum range between low and high tides) and may be exposed to the air for up to ten days between tidal flooding. The upland zone is located just landward of the high marsh, and thus occupies the highest topographical position in the salt marsh system. Although the upland zone is not inundated by tidal waters, the flora in this zone is heavily influenced by its proximity to the salt marsh.

Cordgrass *(Spartina alterniflora)*—1.6 to 10 ft (0.5 to 3 m)

Cordgrass, *Spartina alterniflora*, is a tall coarse grass that grows in the low marsh next to the estuary or tidal creek. This perennial species has a round, hollow stem that is spongy at the base with long (up to 14 in, or 35 cm), smooth leaves that taper to a point. Cordgrass flowers from July through September and has a terminal inflorescence (flowering part) with five to thirty spikes, each containing ten to fifty spikelets. *S. alterniflora* is a pioneer species that can colonize the intertidal zone and initiate marsh development. Thus, cordgrass is extremely tolerant of flooded soil conditions and, once established, it spreads vegetatively by sending rhizomes into surrounding sediments. After it establishes itself in an area, it dominates the landscape and inhibits the growth of other species. Growth forms are tallest along the edge of tidal creeks where the grass is most luxuriant. Stems of *Spartina alterniflora* become progressively shorter towards the high marsh and blend together with *Spartina patens* (salt meadow hay) and *Distichlis spicata* (spike grass).

Salt meadow hay *(Spartina patens)*—1 to 3 ft (30 to 90 cm)

Salt meadow hay, *Spartina patens,* is a finely textured grass that grows landward of, and at a slightly higher elevation than, *S. alterniflora*. It is a perennial plant that has a slender, stiff stem that is hollow. The leaves are very narrow and can be up to 18 in (45 cm) long. *S. patens* also has a terminal inflorescence with three to six spikes alternately arranged. Flowering occurs from late June to October. This species is generally found in irregularly flooded areas of the high marsh and, as the plants age, they usually fall over, forming a mat through which next year's growth will emerge.

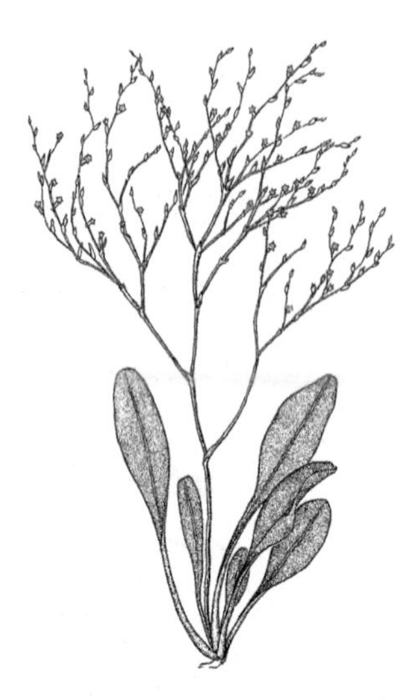

Sea lavender *(Limonium nashii)*—8 to 24 in (20 to 60 cm)

Sea lavender, *Limonium nashii,* is found in the high marsh in salt pannes (shallow, soft-bottomed depressions that retain water during low tides and where salt concentrates). Plants are generally 8 to 23 in (20 to 60 cm) tall and grow in patches. The bases of the stems of sea lavender are thick and wood-like and are surrounded by numerous oval-shaped leaves. These plants possess a stout woody rootstock and a deep taproot and are extremely tolerant of salinities that are greater than that of seawater. In late summer to early fall, sea lavender develops numerous light purple flowers.

Arrow grass (*Triglochin maritimum)*— 8 to 24 in (20 to 60 cm)

Arrow grass, *Triglochin maritimum,* is an erect, fleshy herbaceous plant that grows in low, damp areas in the high marsh. The leaves grow from the base of the plant, are up to 20 in (50 cm) long, and have conspicuous sheaths. Small greenish flowers can be found on the ends of flowering spikes from May through August. The fruits are cup-shaped and surrounded by twelve narrow wings. This species is a pioneer species capable of withstanding regular submergence and extremely low oxygen levels. It is also a common food item of many species of waterfowl.

Solidago sempervirens, commonly known as **seaside goldenrod**, has thick leaves and brilliant golden-yellow flowers in the fall, which makes this attractive marsh plant easy to identify. The leaves are alternately arranged and decrease in size as they approach the top of the plant. In the fall, the flowers are projected above the stem rather than suspended from one side. This species is fairly abundant in the high marsh bordering the adjacent upland.

Seaside goldenrod *(Solidago sempervirens)*—3 to 5 ft (1 to 1.5 m)

Widgeon grass, *Ruppia maritima*, is a submerged aquatic plant that occurs in salt marsh pools or ponds where it is often very abundant. These plants have simple, threadlike leaves with alternately arranged leaf sheaths. Widgeon grass can tolerate moderate to high salinities and thus thrives in salt marsh ponds that are infrequently flooded. In the summer (July to September), flowers and fruits are borne on stalks. Widgeon grass is a prolific seed producer and can colonize large areas fairly quickly. This species is a very important food source for ducks and other waterfowl.

Widgeon grass *(Ruppia maritima)*—up to 39 in (1 meter)

Glasswort or pickleweed *(Salicornia europaea)*—4 to 20 in (10 to 50 cm)

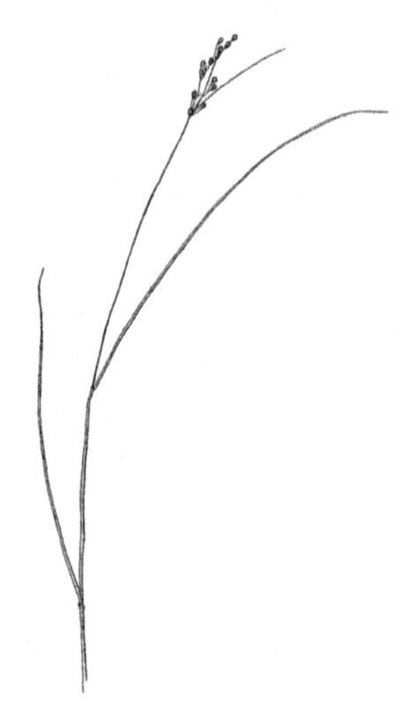

Black rush *(Juncus gerardii)*—8 to 24 in (20 to 60 cm)

Glasswort or **pickleweed**, *Salicornia europaea*, is an annual, pioneer species that is located in the boundary area between the low and high marsh, an area that is frequently subjected to water motion. Due to their branched structure and flexibility of their stems, which offers low frictional resistance, glassworts can survive the strong water currents associated with this area. Glasswort is also extremely tolerant of stresses associated with high salinity, due to its fleshy tissue that conserves moisture (succulence). The blades of the plants are reduced to waxy scales, which wrap around the stem and produce a segmented appearance. Glasswort is usually dark green in the spring and summer but turns yellow, then orange, and then red in the fall.

Black rush, *Juncus gerardii,* stands often form a border on the landward side of grassy marshes. Although grass-like in appearance, *J. gerardii* belongs to a family of plants called rushes. These rushes are quite dark in color, giving rise to the common name for the species, black rush. *J. gerardii* is an erect, herbaceous plant with one or two long leaves and a flower stalk with dense flower clusters near the end of the plant. The fruit capsules usually are dark in color and occur from June to

September. This species tends to have a compact, short, rhizomatous growth (root-like mass spreading horizontally) and is generally found in the high marsh in areas that are irregularly flooded.

Spike grass, *Distichlis spicata,* is a low-growing, fine grass that shares the high marsh with salt meadow hay. Although these two grasses are similar in appearance, they can be distinguished in a number of ways. The stems of spike grass are stiff, hollow, and round with leaves protruding from opposite sides of the stem instead of coming from the base of the plant, as they seem to do in salt meadow hay *(S. patens).* The seeds of *D. spicata* protrude on both sides of rather short heads giving a prickly appearance to the fruiting area (called a spikelet). The flowering period occurs from August to October and the seeds are white rather than colored. The leaves of *S. patens* are also rolled on the edges, more than those of *D. spicata*. Spike grass can be found in clumps around the edges of tide pools, along tidal creeks, and at the high-tide mark. Spike grass is also known to inhabit recently disturbed areas, as well as areas that have considerable salt deposits.

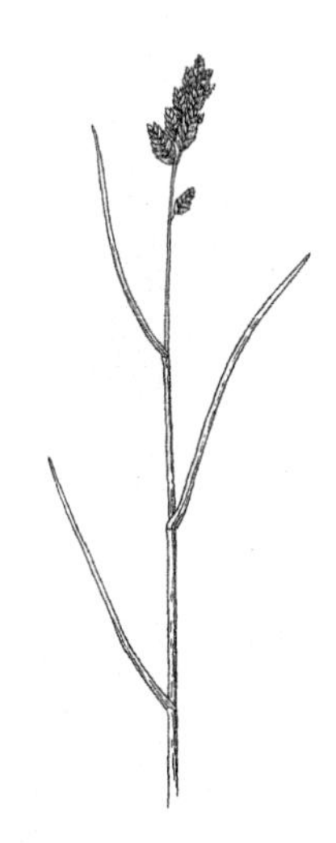

Spike grass *(Distichlis spicata)*—8 to 16 in (20 to 40 cm)

Seaside plantain *(Plantago maritima)*—up to 12 in (30 cm)

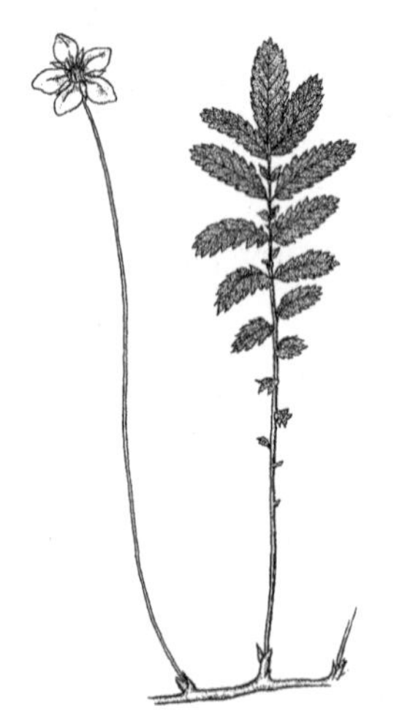

Silverweed *(Potentilla anserina)*—up to 12 in (30 cm)

Seaside plantain, *Plantago maritima,* is a fleshy, herbaceous, low-growing plant comprised of numerous fleshy basal leaves that are narrow and tapered at the ends. Green or whitish flowers are borne on stalks that are generally taller than the basal leaves. Flowering occurs from June to October. Seaside plantain is commonly found in salt pannes (shallow depressions that can retain water at low tide) and irregularly flooded areas in the high marsh.

Silverweed, *Potentilla anserina,* is a low, creeping, herbaceous plant that can reach heights of 12 in (30 cm). It has compound leaves that are sharply toothed. The undersides of *P. anserina* have a silvery appearance, giving the plant its common name, silverweed. It grows on runners and produces bright yellow flowers in the summer. Silverweed is found in irregularly flooded areas of the salt marsh and occasionally on wet sandy beaches.

Zostera marina, or **eelgrass**, primarily occurs subtidally but occasionally extends into the lower intertidal region in areas with sandy or muddy substrate. Eelgrass can be found on the open coast and in sheltered bays, as well as in tidal creeks adjacent to salt marshes. *Z. marina* is a rooted aquatic plant with slender branched stems that have thin, ribbon-like leaves containing 5 to 11 parallel veins. The leaves can be 12 in (30 cm) or more in length and are quite flexible yet strong, due to internal fiber bundles similar to those found in celery.

Eelgrass is the only subtidal vascular plant found in New England. These submerged flowering plants generally grow in a monoculture and are frequently used as "indicator species" to monitor the health of coastal and estuarine communities. Eelgrass beds are also considered critical habitat for a variety of invertebrates, including juvenile scallops.

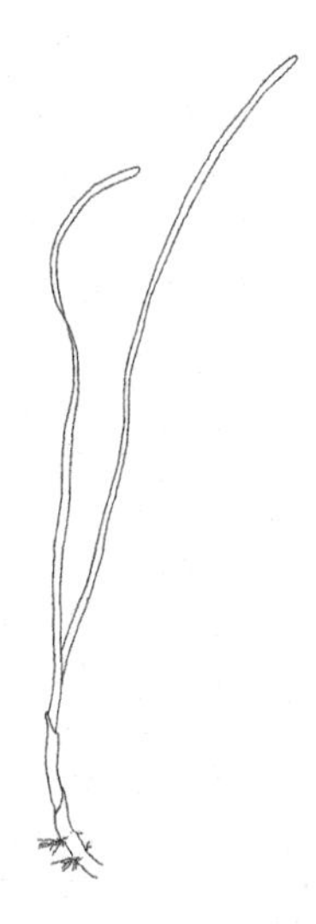

Eelgrass *(Zostera marina)*—generally greater than 12 in (30 cm)

Common Seaweeds (Marine Macroalgae)

Marine macroalgae are beautiful and complex, and the New England coast is an ideal place to discover the great diversity of these interesting and important organisms. To the casual observer walking along the intertidal zone, seaweeds may seem like terrestrial plants that have adapted to life along the shore. However, marine macroalgae are not true plants and differ from them in several ways.

Like terrestrial plants, seaweeds rely on sunlight for photosynthesis. However, seaweeds belong to a more primitive group, called algae, that includes both unicellular microscopic forms (phytoplankton) and multicellular forms, such as seaweeds and certain freshwater algae. Unlike higher plants, seaweed reproduction does not involve flowers and seeds but may involve sexual or asexual spores. Many seaweeds have a complex life history with two or three stages that may be physically very different. Seaweeds come in many diverse forms, ranging from simple species, made of filaments or resembling crusts, to those that are highly complex, such as kelp, which possess specialized tissues and cell types.

As you look at seaweeds along the shore, you will notice that they are found in three basic colors: red, brown, and green. All algae possess chlorophyll, but it is frequently masked by other accessory pigments. These colors are used to divide the seaweeds into three major groups. Green algae are in the division Chlorophyta; red algae are called Rhodophyta; and brown are Phaeophyta. It is important to note that some algae, like the reds, do not always appear red and have several color variations, which resemble brown algae.

If you examine a complex seaweed, such as kelp, you will notice that it has three major parts: holdfast, stipe, and blade. The entire alga is referred to as the thallus. The holdfast acts as an anchor for the seaweed. Despite its resemblance to a root, the holdfast is not used for nutrient uptake. Instead, nutrients are absorbed over the entire thallus. In some species of kelp, the holdfast can be quite large, providing a habitat for many species of small marine invertebrates. Smaller seaweeds have a disc-shaped holdfast, which produces a glue-like substance, allowing the alga to adhere to various substrates.

The seaweed's stipe connects the blade to the holdfast. It is generally rubbery to the touch and is extremely flexible to allow algae to move easily as the tide goes in and out and as waves crash on the intertidal zone.

The blade is the major region of the thallus responsible for photosynthesis, nutrient uptake and, in some kelps, spore production. Kelps possess several distinct types of cells, including sieve tubes that transport nutrients and food from the blade to the major growing region, located between the stipe and the blade. This enables kelp blades in the Northeast to grow up to 8 in (20 cm) per month during the early spring, the height of their growing season. Depending on the structural form of the blade, it can create a refuge for many marine invertebrates, as well as additional substrate for algal and invertebrate epiphytes (organisms that grow on other organisms). It is important to note that not all macroalgae have a prominent holdfast, stipe, and blade as does kelp. Although other species do not have obvious blades, the entire seaweed is still called a thallus.

Seaweed species dominate different zones, or tidal heights, often forming distinct bands on the shore and subtidally. Since species compete for space, light, and nutrients, they settle in zones that best meet their biological requirements and where they will not be out-competed by other species. In addition, wave action, ice scour, desiccation, and predation help determine where seaweeds can survive. Snails, in particular, prey on certain seaweeds in the intertidal zone and help shape seaweed settlement and growth patterns.

When you visit the intertidal zone, you will notice that, at the upper edge of the shore closest to land, there is a yellow crustose growth that covers the rocks. This is not a seaweed but a marine lichen (an organism comprised of one or more species of unicellular algae and a fungus) called *Xanthoria*. Below this lichen is the beginning of the littoral fringe, which is the highest part of the shore regularly reached by splash at spring tides or in rough weather. All of the seaweeds described here occur on the New England coast in the intertidal or shallow subtidal zone.

It should be noted that a few of the seaweeds described have no illustrations because either they closely resemble other seaweeds that are illustrated, or they are so amorphous that a drawing would not adequately capture their shapes. Also, there are some seaweeds that have no common names so only the scientific names are used.

Green Algae (Chlorophyta)

Green algae need a lot of sunlight and are often associated with high nutrient areas. You may find them in lower intertidal and subtidal areas. As long as they receive enough sunlight, these marine algae will thrive. Green algae come in a variety of shapes and sizes; some resemble sheets, while others form long, hollow tubes. Many types of green algae are found in fresh water, but we will focus only on the marine species. Green algae are more closely related to terrestrial plants than are any other algae.

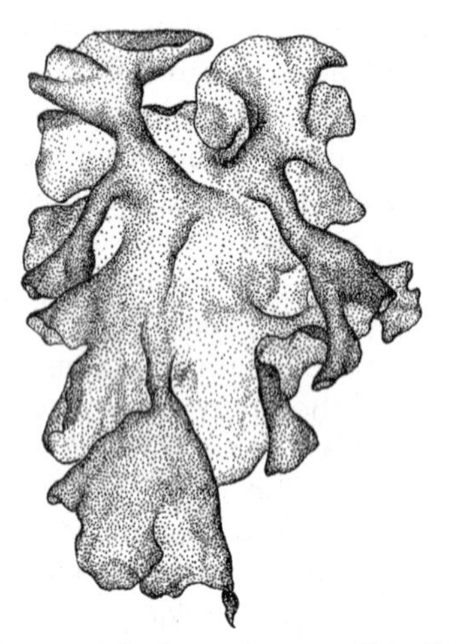

Sea lettuce *(Ulva lactuca)*—up to 20 in (50 cm)

Sea lettuce, *Ulva lactuca,* resembles a thin piece of lettuce, is usually bright green, and is very common year round on the open coast and in estuaries. The alga is only two cells thick and varies in form from circular or oval to narrow and elongated. The edges of the seaweed have waved or ruffled margins and the blade is often perforated with holes of varying sizes that are caused by herbivores, animals that feed on seaweeds or plants. These bright green sheet-like blades are attached by a small, inconspicuous holdfast to hard substrates in the intertidal or in shallow, quiet coves and harbors down to 33 ft (10 m) deep. Sea lettuce is a fast-growing species, often associated with areas of high nutrients, including polluted or recently disturbed habitats where herbivory is low. *Ulva lactuca* closely resembles *Monostroma* (another leafy green alga) except that it has thicker, darker green blades than *Monostroma*, which has only one layer of cells. When *Monostroma* is young, the blade resembles a sac and does not open up into a flat blade until it is older. *Monostroma* is generally found during the spring in mid- to low intertidal tide pools. In many countries, sea lettuce is eaten in soups, salads, and other dishes.

Hollow green weed *(Enteromorpha spp.)*—up to 16 in (40 cm)

Hollow green weed, *Enteromorpha* spp., has a thallus, essentially comprised of several elongated tubes (sometimes constricted), that attaches to the substrate with a small holdfast. The tubes are only one cell thick and their tubular construction helps make the alga buoyant, which is necessary for photosynthesis. The color of *Entero-*

morpha ranges from a bright green, when the seaweed is young, to a darker olive green, when it is old. This species is very common year round on the open coast, as well as in estuaries. Similar to sea lettuce, *Enteromorpha* is often found in areas with high nutrients and may act as an indicator for nutrient enrichment.

Dead man's fingers *(Codium fragile)*—3.3 ft (1 m)

Dead man's fingers, *Codium fragile*, is a very interesting green alga. It is dark green in color and has a spongy texture, due to small internal air sacs that resemble light bulbs under the microscope. Anatomically, *C. fragile* does not contain any cross walls and thus is considered a single-celled organism (also called coenocytic). Interestingly though, it has dichotomous (divided into two parts) branching and can grow to be as tall as 3.3 ft (1 m). This species can be found on the open coast, in estuaries, in the intertidal, and in the subtidal down to 39 ft (12 m). Dead man's fingers is not a native species but was introduced into Maine waters from Europe. This exotic species has now begun to take over a wide range of habitats in the subtidal zone and is reaching nuisance proportion in certain areas. This species is also known as "oyster thief" because it settles on mollusks and quickly becomes densely branched and buoyant enough to tear the mollusks from their beds. Also, *Codium fragile* was once used to pack shellfish for shipping because it stays moist so long.

Chaetomorpha spp. has unbranched filaments, each consisting of a single row of cells. Several different species are found along the New England and Canadian coastline. One resembles a bright green fishing line and is generally tangled among other turf algae in the low intertidal zone. Another species looks like a string of dark green tiny beads. The cells of this species are large enough to see with the naked eye. *Chaetomorpha* is generally found growing in small clumps in the low intertidal zone, as well as in tide pools. This species has a worldwide distribution but, in New England, is more common during the warmer months of the year.

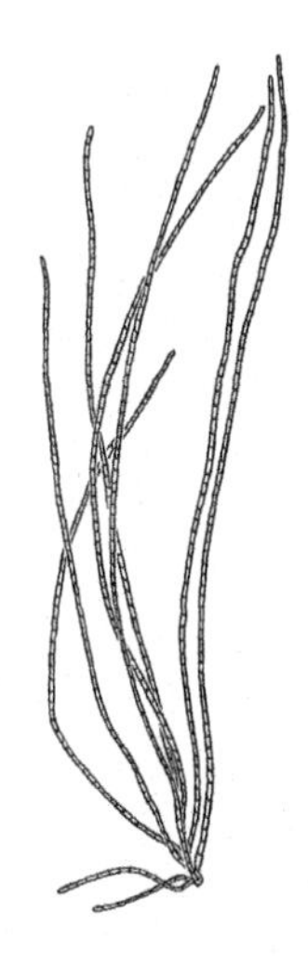

Chaetomorpha spp.—up to 16 in (40 cm)

Another alga that is very similar to *Chaetomorpha* is ***Rhizoclonium tortuosum*** (not shown). *R. tortuosum* is also a tangled, filamentous seaweed that grows in the low intertidal. The filaments are softer and much finer than those of *Chaetomorpha*. *R. tortuosum* is found year-round on the open coast and often forms floating

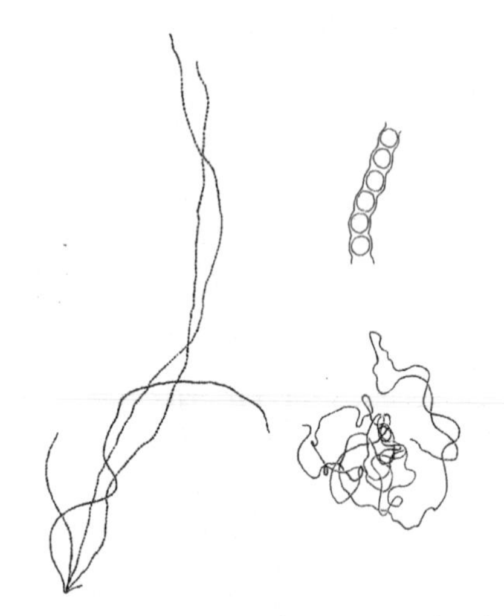

Urospora spp.—greater than 1 cm; detail shows enlarged view

Acrosiphonia arcta and **Spongomorpha aeruginosa**—up to 4 in (10 cm)

mats in estuaries, bays, and salt marshes during the summer.

Urospora spp. is another soft, filamentous green alga. The filament is composed of a chain of large, rounded, swollen cells that are visible to the naked eye. This species is most common in winter and early spring in the intertidal zone on wave-exposed shores. *Urospora* is patchily distributed and often mixed in with other algae.

Some green seaweeds, such as ***Acrosiphonia arcta*** and ***Spongomorpha aeruginosa***, have a very bushy appearance. Both species have relatively long filaments (up to 4 in or 10 cm) that become entangled near the base, causing the branches to take on a rope-like appearance. The rest of the alga has a characteristic ball shape. Several characteristics distinguish the two species, including cell diameter and the presence of root-like structures, called rhizoids, and hooked lateral branches. Both species are fairly common in late winter and early spring in the low intertidal zone and in tide pools.

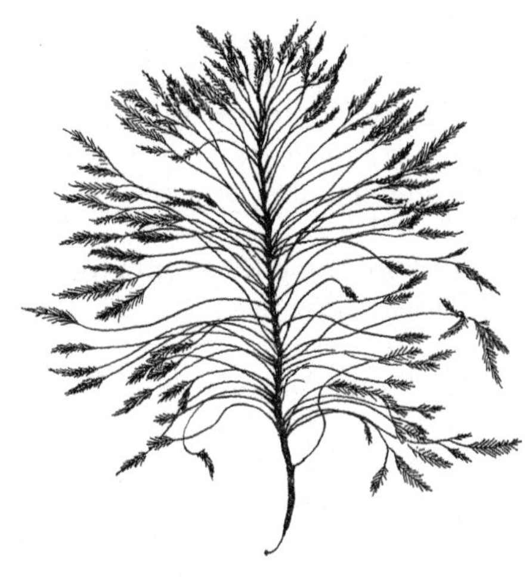

Green sea fern *(Bryopsis plumosa)*—up to 4 in (10 cm)

Green sea fern, *Bryopsis plumosa*, is a short, bushy, light green seaweed. It contains one main thallus with many long, branched, feather-like divergences. This species can be found in tide pools or in protected places on the shore in shallow water. It occurs year-round, but is more common in the spring through fall. *B. plumosa*, like *Codium fragile*, is single-celled.

One last species of green algae that may be encountered in the intertidal zone is ***Ulothrix flacca*** (not shown). This alga consists of small, unbranched filaments. It can be a dominant species in the upper intertidal zone where it completely coats rock surfaces, making walking treacherous. It also occasionally grows epiphytically on the brown seaweeds *Fucus vesiculosus* and *Ascophyllum nodosum* in sheltered areas. This species should not be confused with blue-green algae, which also are found commonly on rocks throughout the intertidal zone and are very slippery to walk on when wet. Blue-green algae are in the division Cyanophyta, which is composed of microscopic, mostly unicellular algae. Blue-green algae have existed for billions of years, and they form colonies that are surrounded by a jelly-like substance, which holds them together and keeps them from drying out. There are many different species of marine Cyanophyta, though all require microscopic examination to determine their species.

Brown Algae (Phaeophyta)
Brown algae are generally easy to locate because they are so large and abundant over a wide area of the intertidal zone. These algae are usually concentrated in the mid-intertidal area to the shallow subtidal. Brown algae provide suitable habitat for many intertidal invertebrates and are extremely important to these organisms' survival. The existence of concentrated stands of brown algae has led to commercial harvesting in many countries where it is used for alginates, organic fertilizers, and fodder.

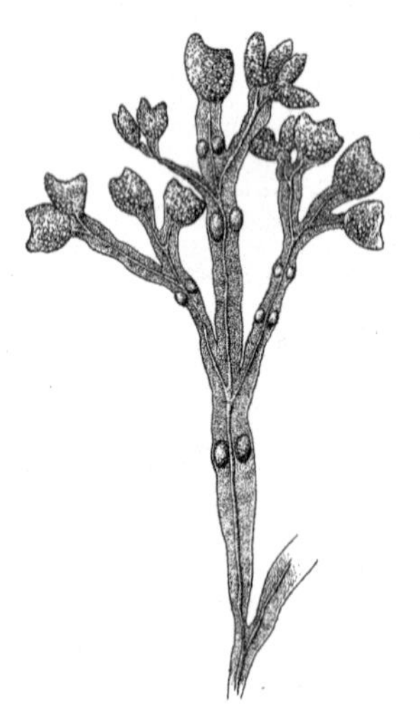

Bladder wrack *(Fucus vesiculosus)*—up to 24 in (60 cm)

The Fucoids
Fucoids have a worldwide distribution. These large brown macroalgae are very abundant throughout the intertidal zone on nearly all rocky shores. Most fucoids contain air bladders, known as vesicles, which help them float above the bottom so they can photosynthesize. They also contain receptacles, inflated bumpy structures, which hold eggs and/or sperm for reproduction. Fucoids range in color from yellowish brown to olive or dark brown and dominate the mid-intertidal zone. *Fucus* spp. and *Ascophyllum nodosum* are the dominant intertidal fucoids in New Hampshire, Maine, and Canada.

There are three different species of *Fucus,* or **bladder wrack**, on the New England coast and several different subspecies. All species in this genus contain receptacles at the tips of their fronds. The shape of the receptacles helps differentiate the different species. *Fucus* is found only in the intertidal zone and is easily recognized by its dichotomous (Y-shaped) branching pattern. The most common species is *Fucus* *vesiculosus,* which usually can be distinguished from the others by its midrib and paired vesicles. However, vesicle development depends on environmental conditions, being more common in seaweeds growing in calm waters. On wave-swept shores, vesicles are often absent and thus cannot always be used to identify the species.

Fucus spiralis receptacle

Another species, ***Fucus spiralis***, can be found in low energy environments and is generally restricted to the upper intertidal. The receptacles of this species are usually swollen, and there is a distinct ridge that surrounds each receptacle. *F. spiralis* is smaller than *F. vesiculosus,* lacks paired air bladders, and has twisted branches.

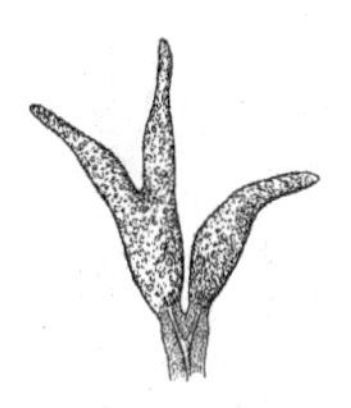

Fucus distichus ssp. ***edentatus*** receptacle

Another common species of *Fucus* that is restricted to exposed shores is ***Fucus distichus*** ssp. (sub species) *edentatus*. This fucoid has elongated, spear-shaped receptacles, lacks vesicles, and grows in the low intertidal. A related species also found on Maine shores is *F. distichus* ssp. *distichus,* which is much smaller than *F. distichus* ssp. *edentatus,* is only found in high tide pools, and contains no air bladders or noticeable receptacles.

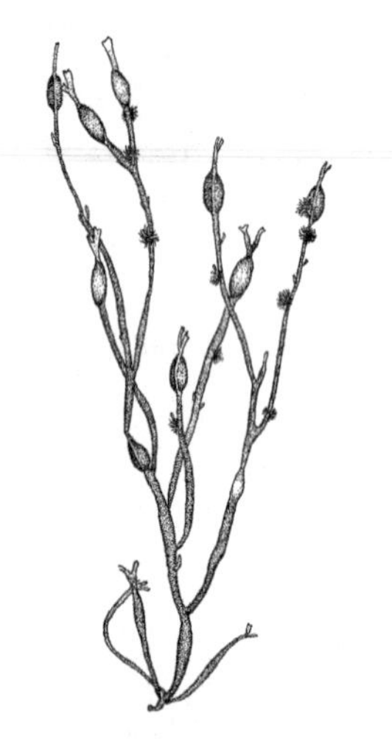

Knotted wrack or rockweed *(Ascophyllum nodosum)*—20 to 29 in (50 to 75 cm) tall but can reach 10 ft (3 m) in length on sheltered shores; (shown with ***Elachista fucicola*** as an epiphyte)

Another common intertidal alga in the North Atlantic is **knotted wrack**, *Ascophyllum nodosum*. This species is olive green in color and usually 20 to 29 in (50 to 75 cm) tall but can reach 10 ft (3 m) in length on sheltered shores. Main axes of the thallus are compressed and branched in two parts, or dichotomously. All except the oldest parts of the thallus bear numerous, short, secondary lateral branches. At intervals along the axes, prominent air bladders are present that provide buoyancy to support the thallus while submerged. The air bladders are formed annually and thus serve as a means of aging the seaweed. *A. nodosum* commonly has a life span of 8 to 10 years but can grow to be much older (20 to 23 years). This species is reproductive in the spring (May to June), producing small receptacles attached by short stalks along the length of the thallus. The receptacles become very swollen when ripe and, following release of the sperm or eggs, they will drop off the seaweed. Easy access to large quantities of this species has led to commercial harvesting of *Ascophyllum nodosum* in many countries where it is used for emulsifiers and thickening agents, organic fertilizers, and animal fodder.

The Kelps

The *Laminariales*, commonly called kelps, dominate the lower intertidal and subtidal zones in temperate to Artic latitudes in the Northern Hemisphere. This group includes the largest and most complex of the brown algae. **Sugar kelp**, *Laminaria saccharina*, and *Laminaria longicruris* (not shown) are both very common kelps on the coast of Maine. In recent years, kelp abundance has increased in this region, partly due to the commercial harvest of sea urchins, the primary grazers on kelp.

Sugar kelp *(Laminaria saccharina)*—up to 10 ft (3 m)

L. saccharina is a shallow subtidal species that can reach 10 ft (3 m) in length. This species also can be found readily in tide pools. It has a blade; a short, cylindrical stipe; and a holdfast with finger-like extensions, called haptera. All kelp species are dark brown and smooth and slippery to the touch. Depending on the environmental conditions, the blade can be either very flat or have ruffled edges. You will often see lacy bryozoans (small colonial invertebrates) living on the surface of kelp blades. The bryozoans resemble a network of small white boxes.

Laminaria longicruris (not shown) is another kelp species found on the coast of Maine; however, it is restricted to the subtidal. The only time you will encounter this species will be as drift algae. The difference between the two *Laminaria* species is that *L. longicruris* has a longer, hollow stipe, whereas *L. saccharina's* stipe is shorter and solid. Also, *L. longicruris* has a wider and much longer blade than does sugar kelp.

Horsetail kelp *(Laminaria digitata)*—up to 20 in (50 cm)

Horsetail kelp, *Laminaria digitata,* is easily identified by the many tail-like sections that make up the alga, which explains its common name. The blade can be divided into as many as 30 different segments. This structure reduces drag on the seaweed, allowing *L. digitata* to live and thrive in high-energy environments in low intertidal and subtidal zones. The stipe is relatively short and flattens near the base of the blade.

Winged kelp *(Alaria esculenta)*—up to 3 ft (90 cm)

Winged kelp, *Alaria esculenta,* has several noteworthy features. First, it has a very prominent flat midrib that runs the length of the blade. Secondly, it has a set of leaf-like structures, called sporophylls, which are attached to the stipe below the blade. These sporophylls are the reproductive structures of the organism. *A. esculenta* is a perennial species that occurs in tide pools and in the shallow subtidal zone on wave-swept, high-energy shores.

The blades are fairly narrow and often have a fringed appearance, due to the blade splitting from wave energy.

Sea colander or **shotgun kelp**, *Agarum clathratum,* is another type of kelp that looks like Swiss cheese or a colander. The blade is fairly coarse in texture, is undivided, contains a well-defined midrib, and has numerous holes throughout. *A. clathratum* is found in the low intertidal zone, and looks as if it has been eaten by an herbivore. The holes are part of the blade and should not be confused with damaged pieces of *Laminaria* spp. Unlike *A. esculenta* and *L. digitata,* this species likes to grow in low energy areas.

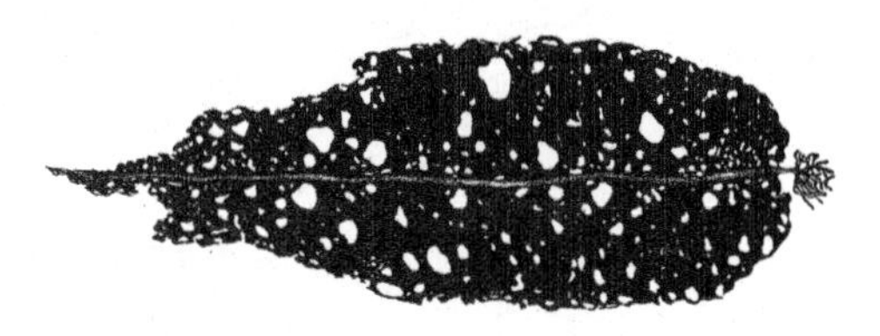

Sea colander or shotgun kelp *(Agarum clathratum)*—up to 18 in (45 cm)

Other Brown Algae

The common names of *Leathesia difformis,* also known as **sea potato** or **rat's brain**, are very good descriptions of its appearance. *L. difformis* is a hollow, yellowish brown, globular mass approximately 0.8 to 1.2 in (2 to 3 cm) in width. This small alga grows on top of other algal species in the mid- to low intertidal. When you first encounter this species, it looks like a growth on the algae, but if you remove it, you will see that it is a hollow sac. It is sometimes confused with an invertebrate egg sac.

Sea potato or rat's brain *(Leathesia difformis)*—0.8 to 1.2 in (2 to 3 cm) in width

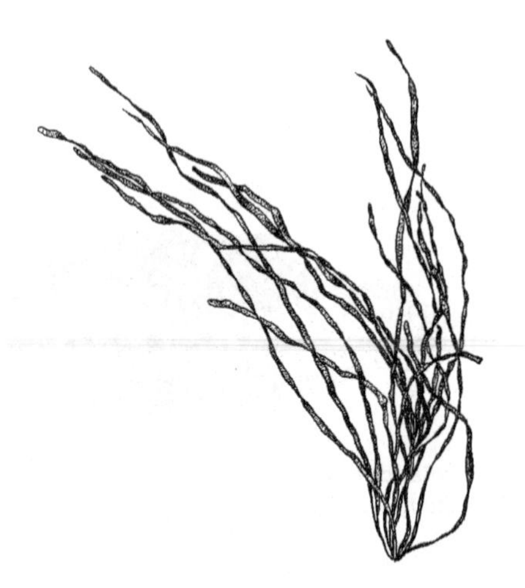

Sausage weed *(Scytosiphon simplicissimus)*—up to 24 in (60 cm) but is generally 6 to 10 in (15 to 25 cm) in length

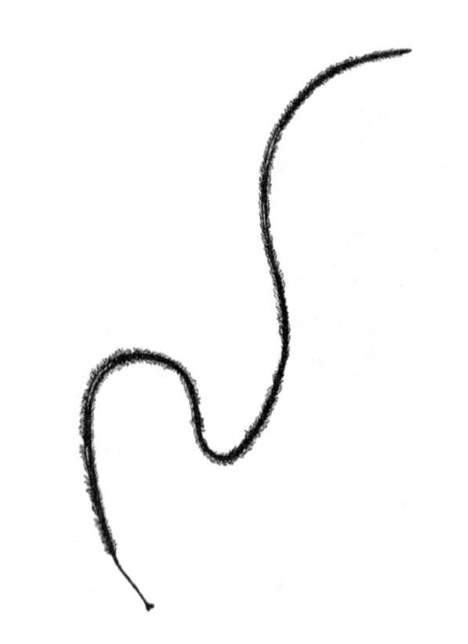

Cord weed *(Chorda tomentosa)*—less than 3 ft (90 cm) long

Sausage weed, *Scytosiphon simplicissimus,* has a hollow, tubular thallus that often has regular, sausage-like constrictions. This thallus alternates with an encrusting phase resembling *Ralfsia* (see p. 92). Different environmental conditions influence the seasonal occurrence of the upright versus the encrusting form. Short daylight periods are believed to produce the upright thallus, whereas long daylight hours produce the crustose phase. *S. simplicissimus* has a short stalk, a disc-like holdfast, and unbranched hollow tubes. It can grow to be quite tall (24 in or 60 cm) but is generally 6 to 10 in (15 to 25 cm) in length in this region. It tends to grow in clusters and can be found on shells, stones, and rock surfaces.

Cord weed, *Chorda tomentosa,* has an interesting form with a thallus that is unbranched and whip-like. The alga's entire surface is densely covered with delicate, colorless hairs. *C. tomentosa* is considered to have a primitive morphological condition because the thallus does not differentiate into a blade and stipe, and the holdfast is disc-like without any finger-like extensions, or haptera. It is a late winter to early spring annual and occurs commonly near mean low water. Cord weed can also be found attached to docks and pilings.

Black whip weed, *Chordaria flagelliformis,* is a dark brown seaweed with whip-like branches that project from a single central axis. Both the main axis and the branches are slippery and cord-like. The whole seaweed tapers near the base. A disc-like holdfast attaches *C. flagelliformis* to rocks in the lower intertidal and in tide pools.

Black whip weed *(Chordaria flagelliformis)*—up to 2 ft (60 cm)

Soft sour weed, *Desmarestia viridis,* is light to dark brown, and it looks soft and flowing when in the water. It has opposite branches that extend from a cylindrical main axis. Soft sour weed is most abundant in spring and early summer and, as its common name suggests, it has a very sour odor. *D. viridis* produces sulfuric acid inside its cells and, when removed from the water, the alga begins to degenerate. Apparently the cells break down and release the acid, which is highly corrosive, so this species should be isolated from the rest of the collection.

Soft sour weed *(Desmarestia viridis)*—1 to 2 ft (30 to 60 cm)

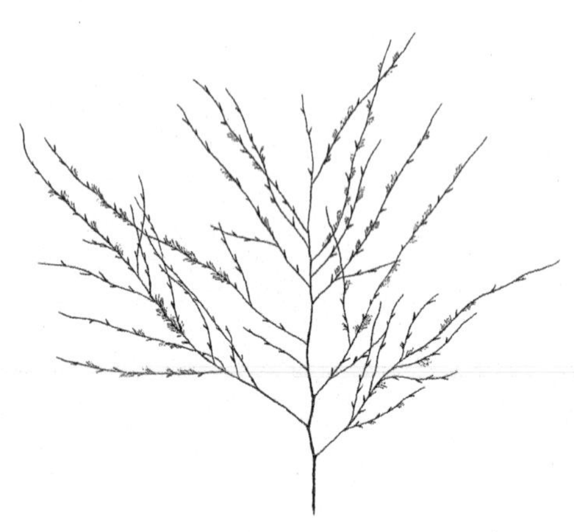

Spiny sour weed *(Desmarestia aculeata)*—usually 1 to 2 ft (30 to 60 cm) but can be longer

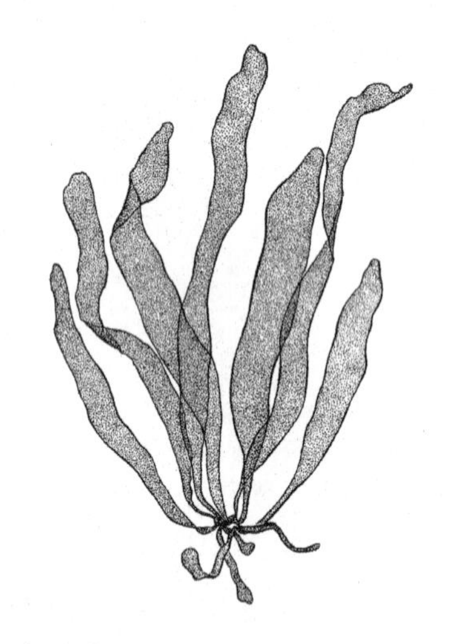

Petalonia fascia—up to 1 in (2.5 cm) wide and 8 in (20 cm) long

Spiny sour weed, *Desmarestia aculeata,* is another related brown alga. This species is finely divided with an alternate branching pattern. Unlike *D. viridis,* which is very soft, this species of *Desmarestia* has a stiff, wiry thallus with small spiny projections. Both species of *Desmarestia* are subtidal but, occasionally, they are found as drift algae.

Petalonia fascia consists of flat, ribbon-like blades up to 1 in (2.5 cm) wide and 8 in (20 cm) long. The blade tapers sharply, often asymmetrically, to a cylindrical stipe, which is attached to a flat, circular holdfast at the base. *P. fascia* occurs in the littoral zone in shallow tide pools during winter and early spring.

Two of the most likely filamentous brown seaweeds that you will encounter in the intertidal are ***Pilayella littoralis*** and ***Ectocarpus***. *P. littoralis* is a filamentous, densely branched alga that can grow to 6 in (15 cm) in length. It is a yellowish brown epiphyte (an organism growing on another organism) generally found on knotted wrack and bladder wrack although it occasionally attaches to rocks or is found free-floating. It is extremely abundant on sheltered shores in the spring and fall. Often the filaments become twisted, forming visible, rope-like strands.

Pilayella littoralis—up to 6 in (15 cm)

Ectocarpus spp. is another very similar filamentous brown alga. It also produces yellowish brown tufts that can be found growing epiphytically on other seaweeds, attached to rocks, or free-floating. This species is generally much longer (up to 20 in, or 50 cm) than *Pilayella,* and it is less abundant. These two species are often difficult to distinguish in the field.

Ectocarpus spp.—up to 20 in (50 cm)

Elachista fucicola—0.4 to 0.6 in (1 to1.5 cm)

Elachista fucicola is also a common brown epiphyte. This seaweed forms short, dense tufts that arise from a basal cushion, part of which penetrates the host tissue. The erect filaments are 0.4 to 0.6 in (1 to 1.5 cm) tall and densely branched. This species, generally found only on *Ascophyllum nodosum* and *Fucus vesiculosus*, is most common during the summer months.

The most abundant crust-forming brown seaweeds are species of ***Ralfsia***. There are two different species of *Ralfsia* (*R. fungiformis* and *R. verrucosa*) in Maine. *Ralfsia* spp. can be found in the intertidal to the subtidal zone and is abundant on rocks, shells, and driftwood where it forms thin discs up to 0.8 in (20 mm) in diameter. The margins of the brown crust show a banding pattern with overlapping lobes. The crust is generally circular and smooth when the specimen is young, but it becomes irregularly shaped, rough, and brittle with age.

Ralfsia spp.—thin discs up to 0.8 in (2 cm) in diameter

Red Algae (Rhodophyta)

Red algae are the most numerous of the marine macroalgae and exhibit an incredible range of form and lifestyle. Some are branching with many divisions, others are filamentous, and still others are encrusting. There are annuals and perennials, free-living species, and parasites. There are more species of red seaweeds than all other seaweeds combined.

Physiologically, red algae require less sunlight to grow and proliferate than do green algae and thus can survive in the intertidal in shaded habitats. Many red algal species are adapted to low light environments and cannot survive in full sunlight. As a result, most red algae are hidden under the dominant fucoid canopy and in crevices and tide pools where light levels are reduced. Red algae are often used as a direct food source and for commercial colloidal extracts (e.g., agar and carrageenan). Carrageenan is odorless, tasteless, and has no color, which makes it ideal to use in recipes as a thickening agent.

Irish moss *(Chondrus crispus)*—4 to 6 in (10 to 15 cm)

Irish moss, *Chondrus crispus,* is a common low intertidal red alga that can extend subtidally to 230 ft (70 m) deep. It has a disc-like holdfast that supports a flattened, dichotomously branched thallus. This species is dark reddish brown to purple and has blunt or rounded tips. It is very abundant and usually grows in large populations on most exposed rocky shores in the North Atlantic. *C. crispus* can have several different forms, depending on wave action, population density, and grazer intensity. This alga is commercially harvested for its carrageenan.

False Irish moss *(Mastocarpus stellatus)*—4 to 6 in (10 to 15 cm)

Within and adjacent to *C. crispus* beds, another red alga, *Mastocarpus stellatus*, can be found. This alga looks very similar to Irish moss, which is why the common name for this species is **false Irish moss**. The only differences between the two species are that *M. stellatus* has slightly curled blades with small bumps on the underside and *Chondrus crispus* does not. Because false Irish moss is more tolerant to freezing than is Irish moss, it is found slightly higher in the low intertidal zone.

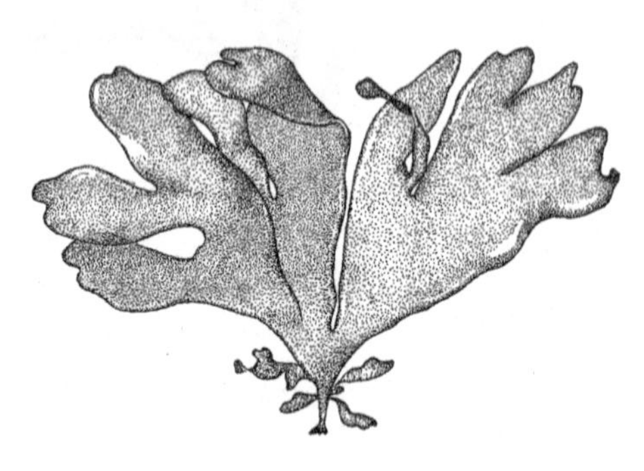

Dulse *(Palmaria palmata)*—up to 20 in (50 cm) in length

Dulse, *Palmaria palmata*, has flat, broad blades that are usually dichotomously divided. This species is purplish red in color and quite leathery. It often has small proliferations or outgrowths along the margin, particularly toward the base where there is a small disc-like holdfast. *P. palmata* is found in the mid-to low intertidal zone where it can reach up to 20 in (50 cm) in length. This species becomes more abundant as you proceed east towards the Canadian border. Dulse is edible and considered a delicacy.

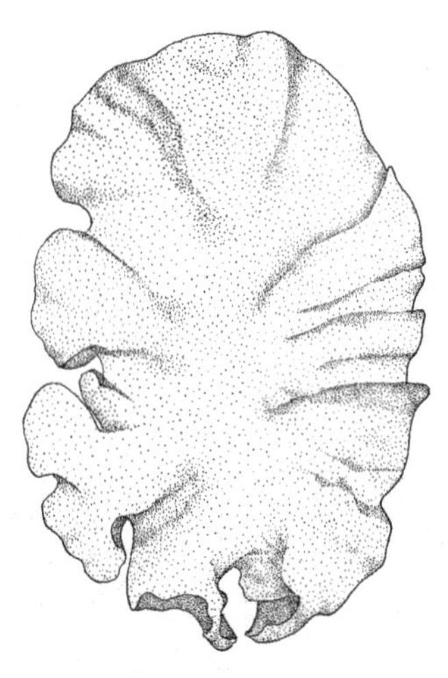

Nori or laver *(Porphyra* spp.*)*—usually 12 in (30 cm) but can be longer

Another edible red seaweed found in the intertidal is *Porphyra* spp., commonly called **nori** or **laver**. You may recognize the name of this alga from Japanese cuisine where it is used as a wrapping for sushi, a roll made of pressed rice garnished with fish and vegetables. Nori is a thin, soft, sheet-like seaweed with a very small disc-like holdfast. Most species in this genus are perennial, with the leafy stage appearing in late November and disappearing about mid-April. Small, basal portions of the leafy alga persist and give rise to new upper parts the following season. Several different species are found along the coast and can only be taxonomically separated based on microscopic features. *P. umbilicalis* is the most abundant species in the area, growing on rocks and wharves in the low intertidal zone. It is purple to purplish brown in color.

Sea oak *(Phycodrys rubens)*—up to 8 in (20 cm) long

Phycodrys rubens, commonly called **sea oak**, is a subtidal species that is occasionally found in deep tide pools in the intertidal zone.

P. rubens has a delicate, pinkish red blade, with broad leaves that contains a distinct midrib and lateral veins. The veins resemble the vein systems in the leaves of oaks, hence its name. The blades arise from a solid stem and frequently bear small proliferations at the blade margins. The leafy blades are deeply lobed and toothed at the edges. This species is common year round on the open coast and in the subtidal zone in estuaries.

Membranoptera alata—less than 4 in (10 cm)

Another red alga that is similar to, but smaller than, *P. rubens* is ***Membranoptera alata***. This species also contains a prominent midrib but has very faint lateral veins. The blade is deep crimson in color and less than 4 in (10 cm) high. Unlike *P. rubens*, which has a lobed blade, *M. alata* has a narrow, alternately branched, flattened blade. This alga is commonly found as an epiphyte on coarser seaweeds in the subtidal zone.

Red fern or feather weed *(Ptilota serrata)*—can reach 6 in (15 cm) high

Red fern or **feather weed**, *Ptilota serrata,* is a red seaweed that becomes more conspicuous in the shallow subtidal zone as one heads northward. The alga is stiff, bushy, purplish red in color, and can reach 6 in (15 cm) high. It also has a feather-like branching pattern that occurs in one plane. *P. serrata* is common in the subtidal, attached to the base of kelps or on rocks, but it also is found frequently in the intertidal zone as drift algae.

Plumaria plumosa is another finely feathered red alga. Like *Ptilota*, it prefers to grow in the subtidal zone, attached to the base of larger macroalgae. However, it can be found in the intertidal zone growing on vertical rock surfaces under macroalgal canopies. *P. plumosa* branches are flattened, feather-like, and dark brown to reddish purple in color.

Plumaria plumosa—up to 4 in (10 cm)

The defining characteristic of the red alga **Rhodomela confervoides** is the flattened tufts at the ends of the branches. The thallus is wiry at the base and then becomes quite bushy towards the top of the seaweed. The hair-like clusters of branches fall off during the summer, making this species more difficult to identify. *R. confervoides* can be found year-round in tide pools and in the subtidal zone.

Rhodomela confervoides—up to 16 in (40 cm)

Callophyllis cristata—2 to 6 in (5 to15 cm)

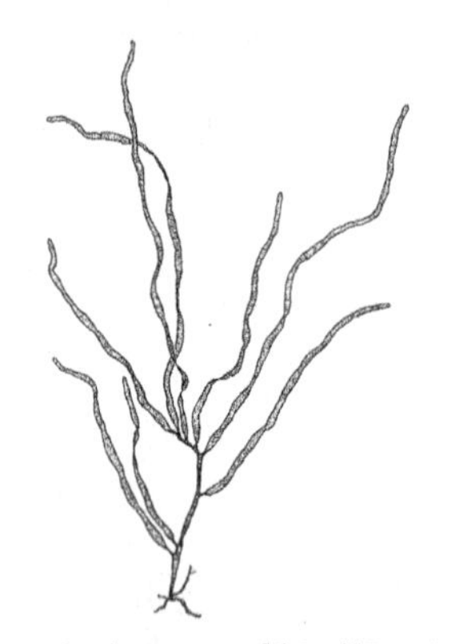

Dumontia contorta—up to 20 in (50 cm) tall

Develaraea ramentacea—up to 16 in (40 cm)

Callophyllis cristata is a small, bushy, attractive rose-pink subtidal alga. The repeatedly branched, flat, spreading axes grow to a height of to 2 to 6 in (5 to 15 cm) from a small basal disc. It is interesting to note that the alga decreases in size each time the branches divide. *C. cristata* grows on mussels and sponges, or it attaches to the bases of large kelps. It is typically found in deeper water but also may be found in tide pools.

Dumontia contorta is a seasonal alga, generally found in shallow tide pools during the spring and early summer. It has long, hollow branches that are often twisted. *D. contorta* can grow to 20 in (50 cm) tall and has a dull reddish color that becomes yellowish when exposed to bright light. This species is often confused with ***Develaraea ramentacea.***

D. ramentacea can be very similar in appearance to *D. contorta*, depending on the environmental conditions. This species is extremely variable in form, ranging from unbranched tubular fronds arising from a small holdfast, to heavily branched axes, to a single axis with numerous, short proliferations. In Maine, this species has a winter and summer form, both of which display somewhat different structures.

D. ramentacea can be identified based on the stiffness of the main axes. *D. contorta* is much longer and quite often twisted.

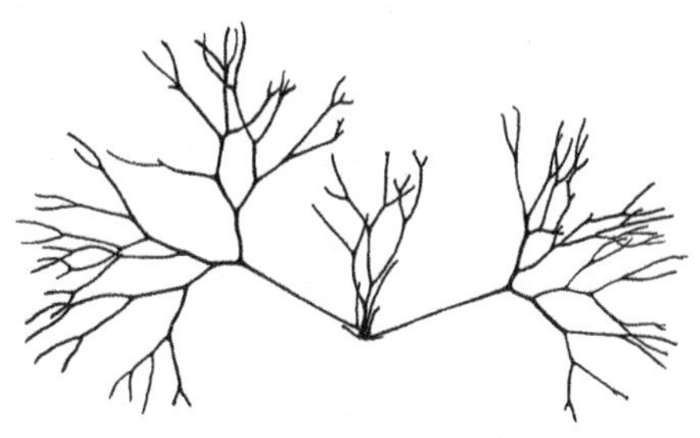

Polyides rotundus—4 to 5 in (10 to 13 cm) high

Another stiff red alga that is cylindrical is ***Polyides rotundus***. This species grows as several erect, round stalks arising from a single disc-like holdfast. The branches divide six to eight times in a regular dichotomous pattern while remaining the same thickness. This species appears to be black in color and is found year-round in low intertidal tide pools.

Agardhiella subulata is a bushy seaweed with relatively slender branches that taper at both ends. The main axes are cylindrical like *Polyides* though not as thick. *Agardhiella* can grow to be 12 in (30 cm) long and prefers relatively quiet water. It can be found occasionally in the low intertidal zone attached to stones or shells. This species is near its northern limit in Maine.

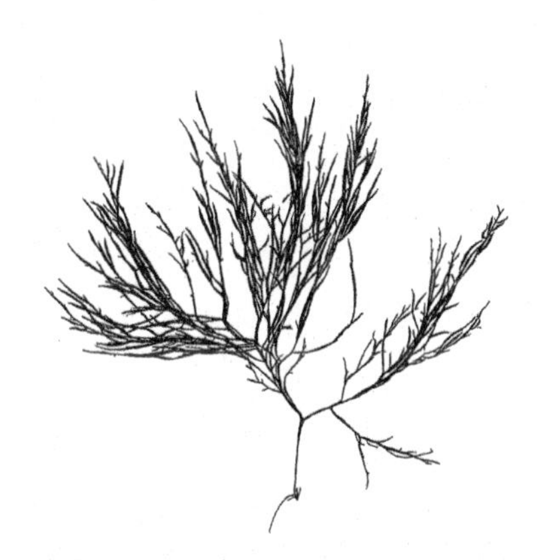

Agardhiella subulata—12 in (30 cm) long

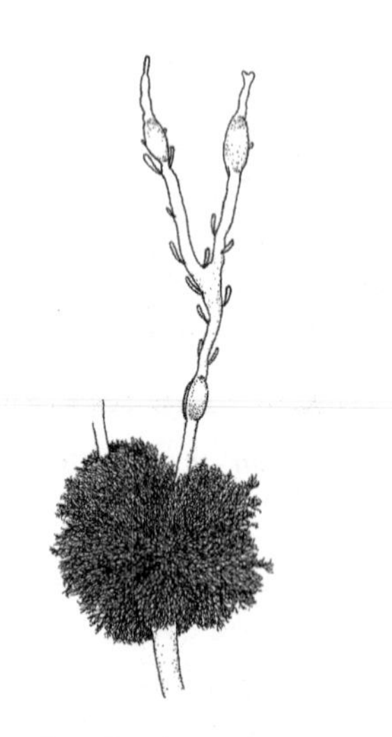

Tubed weed (*Polysiphonia lanosa*)—1 to 3 in (3 to 8 cm) tall; shown growing as epiphyte on knotted wrack

Ceramium spp.—6 in (15 cm)

Tubed weed, *Polysiphonia lanosa,* grows as an epiphyte attached to the brown alga, *Ascophyllum nodosum*. Individuals are 1 to 3 in (3 to 8 cm) tall, blackish purple in color, with soft erect filaments. This species is profusely branched and appears as black tufts. Tubed weed is found only on wave-exposed coasts and is virtually absent in sheltered areas.

Ceramium spp. is a rather coarse seaweed with tufts of dichotomously branched filaments, which have distinctive pincers at the tips. Most individuals are erect and bushy, though sometimes they are creeping. The cells of the main axes are corticated (having a secondary layer of small surface cells) at the nodes by a zone of smaller cells, whose arrangement is important in identifying species within the genus. This cortication is usually visible to the unaided eye; however, it is impossible to identify most *Ceramium* to the species level without microscopic examination. This genus is widespread and can be found in the lower intertidal on rocks, in tide pools, as epiphytes on other seaweeds, and on almost any firm substrate.

One of the most interesting groups of red algae is that which incorporates calcium carbonate into its cellular structure. Seaweeds of this type come in two basic forms: upright and crustose. *Corallina officinalis*, or ***coral weed***, is a very common coralline alga that has an upright, calcareous thallus with jointed segments. This species generally grows in tufts, is usually 2 to 3 in (5 to 8 cm) tall, and has flattened branches arranged on opposite sides of the axis. *C. officinalis* can be found in tidal pools, as well as in the shallow subtidal zone. It is the only upright calcareous species in this region.

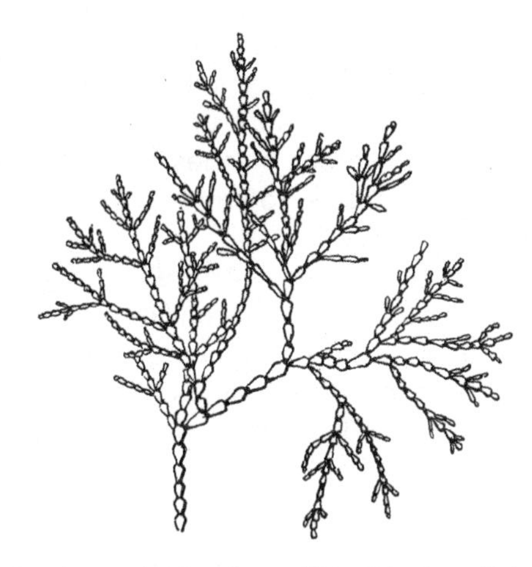

Coral weed (*Corallina officinalis*)—2 to 3 in (5 to 8 cm) tall

The New England coast is also home to several different species of crustose coralline algae. ***Lithothamnion glaciale*** is a pink crustose coralline that produces a thick calcium layer over rocks and shells. It has a knobby, rough texture and often completely covers mussel shells and small rocks. The thallus is usually pale to dark pink with whitish tips. *Phymatolithon* spp. (not shown) is another crustose coralline alga commonly found in the intertidal zone in tide pools and under rockweed canopies. This alga is a very thin, rose-colored crust with coarse ridges. The smoothest of the corallines found in this region is

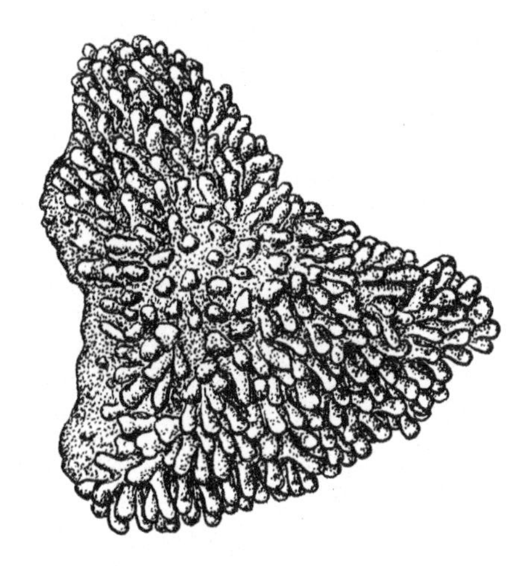

Lithothamnion glaciale—0.8 in (2 cm) thick

Clathromorphum spp. (not shown), which has an irregular, overlapping shape and can be up to 2 in (5 cm) thick. The color varies from pale to purplish pink, to tan or white.

Some red algae have a crustose form but are not considered corallines. These species produce a fleshy crust that clings to rocks very tightly. *Hildenbrandia rubra* (not shown) is the most abundant noncalcified encrusting red algae found on the surfaces of pebbles, rocks, and shells in the intertidal zone. The crusts vary in color from brownish red to purplish red, and they may be up to 0.02 in (0.5 mm) thick. This species is principally intertidal and forms extensive beds under algal canopies.

INDEX

Contributors

Les Watling
Les is a professor of oceanography and marine biology in the University of Maine's School of Marine Sciences. In 1998, he was honored as a Pew Fellow in Marine Conservation. He received his Ph.D. in biological oceanography at the University of Delaware in 1974 and joined the University of Maine as a faculty member in 1976. His office and research lab are located at the University's Darling Marine Center in Walpole.

Jill Fegley
Jill received her Ph.D. in ecology and environmental science from the University of Maine. Her dissertation research on seaweed harvesting was awarded the National Ocean Service, Walter B. Jones Award for Excellence in Coastal and Marine Graduate Study. She is currently an assistant professor of marine biology at Maine Maritime Academy in Castine, where she teaches marine botany, marine biology, and general biology.

John Moring
John received his Ph.D. in fisheries from the University of Washington in 1973. After graduation, John worked for the Oregon Department of Fish and Game as a research biologist. He then moved to Maine in 1979 to assume the position of assistant leader for fisheries of the Maine Cooperative Fish and Wildlife Research Unit, with parallel appointments as professor of zoology and marine sciences at the University of Maine, positions he held until his death on May 9, 2002. He had published 159 scientific papers and dozens of reports, summaries, and abstracts. John was an accomplished freelance writer, publishing more than sixty articles, humor pieces, short stories, poems, and two nonfiction books.

ANDREA SULZER

"... it is nicer to think than to do, to feel than to think, but nicest of all merely to look." —Goethe.

Andrea says that drawing helps her learn how to look—how to pay attention. She was the illustrator for the first edition of this field guide, and has illustrated *Maintaining Biodiversity in Forest Ecosystems, The Whole Paddler's Catalog,* and *Fundamentals of Conservation Biology.*

SUSAN WHITE

Susan is the assistant director and communications coordinator for the Maine Sea Grant Program. She received a B.A. in psychology from Earlham College and, more recently, was involved in a graduate program in marine biology at Northeastern University. Prior to her tenure at Sea Grant, Susan worked as a video/ multi-image producer and taught language arts, drama, and art at an alternative school. While at Sea Grant, Susan has written and edited several books, including *A Lobster in Every Pot: Recipes and Lore, A Field Guide to Economically Important Seaweeds of Northern New England,* and *A Guide to Common Marine Organisms Along the Coast of Maine.*

MAINE SEA GRANT supports marine science research and outreach activities to promote the understanding, sustainable use, and conservation of ocean and coastal resources. Based at the University of Maine in Orono, the statewide program works in partnership with marine industries, scientists, government agencies, private organizations, and a wide range of marine resource users. Maine Sea Grant is part of the National Sea Grant College Program, a network made up of thirty programs located in the coastal and Great Lakes states. Sea Grant is a partnership between the nation's universities and the National Oceanic and Atmospheric Administration.
www.seagrant.umaine.edu

CPSIA information can be obtained
at www.ICGtesting.com
Printed in the USA
BVOW08s2238061017
496817BV00002B/2/P